PRELIMINARY DISCOURSE
TO THE ENCYCLOPEDIA
OF DIDEROT

The Library of Liberal Arts
OSKAR PIEST, FOUNDER

PRELIMINARY DISCOURSE TO THE ENCYCLOPEDIA OF DIDEROT

JEAN LE ROND D'ALEMBERT

Translated by
RICHARD N. SCHWAB

with the collaboration of
WALTER E. REX

with an introduction and notes by
RICHARD N. SCHWAB

The Library of Liberal Arts
published by
Bobbs-Merrill Educational Publishing
Indianapolis

Jean Le Rond d'Alembert: 1717-1783

PRELIMINARY DISCOURSE TO THE ENCYCLOPEDIA OF DIDEROT was originally published in 1751

• • • • • • • • • • • • • • • • • • • •

Copyright © 1963 by The Bobbs-Merrill Company, Inc.
Printed in the United States of America
All rights reserved. No part of this book shall be reproduced or transmitted in any form or by any means, electronic or mechanical, including photocopying, recording, or by any information or retrieval system, without written permission from the Publisher:

       The Bobbs-Merrill Company, Inc.
       4300 West 62nd Street
       Indianapolis, Indiana 46268

First Edition
Sixth Printing—1980
Library of Congress Catalog Card Number: 63-21831
ISBN 0-672-60276-8 (pbk.)
ISBN 0-672-51037-5

Frontispiece by C.-N. Cochin fils, *appearing in the 1751 edition of the* Encyclopedia.

Beneath a temple of Ionic architecture, sanctuary of Truth, we see Truth wrapped in a veil, radiant with a light which parts the clouds and disperses them.

On the right of Truth, Reason and Philosophy are engaged, the one in lifting the veil from Truth, the other in pulling it away.

At her feet Theology, on her knees, receives her light from on high.

Following the line of figures, we see grouped on the same side Memory, and Ancient and Modern History; History is writing the annals, and Time serves as a support for her.

Grouped below are Geometry, Astronomy, and Physics.

The figures below this group show Optics, Botany, Chemistry, and Agriculture.

At the bottom are several Arts and Professions that proceed from the Sciences.

On the left of Truth we see Imagination, who is preparing to adorn and crown Truth.

Beneath Imagination, the Artist has placed the different genres of Poetry—Epic, Dramatic, Satiric, and Pastoral.

Next come the other Arts of Imitation—Music, Painting, Sculpture, and Architecture.

(*Translated from the 1751 edition of the* Encyclopedia.)

CONTENTS

LIST OF ILLUSTRATIONS	viii
TRANSLATOR'S INTRODUCTION	ix
SELECTED BIBLIOGRAPHY	li
NOTE ON THE TRANSLATION	lv

PRELIMINARY DISCOURSE TO THE ENCYCLOPEDIA OF DIDEROT

Part I	3
Part II	60
Part III	106
DETAILED EXPLANATION OF THE SYSTEM OF HUMAN KNOWLEDGE	143
OBSERVATIONS ON BACON'S DIVISION OF THE SCIENCES	159
INDEX	165

ILLUSTRATIONS

Frontispiece of the *Encyclopedia*   v

Detailed System of Human Knowledge   144

INTRODUCTION

> In a word, Madame, I can assure you that while writing this work, I had posterity before my eyes at every line.—
> *d'Alembert to Madame du Deffand, December 22, 1752.*

Of all the shorter works of the eighteenth-century *philosophes*, the *Preliminary Discourse* to Diderot's *Encyclopedia* is incomparably the best introduction to the French Enlightenment.[1] It *is* the Enlightenment insofar as one can make such a claim for any single work; with a notable economy and vigor it expresses the hopes, the dogmas, the assumptions, and the prejudices we have come to associate with the movement of the *philosophes*. From the moment of its publication in 1751, many leaders of the Enlightenment recognized it as a masterful statement of their "philosophy," and even the men who were fearful of its implications acclaimed it for its lucidity and compactness. No less a judge than the great Montesquieu complimented d'Alembert on his work in the most flattering terms: "You have given me great pleasure. I have read and reread your *Preliminary Discourse*. It has strength, it has charm, it has precision; richer in thoughts than in words, likewise rich in sentiment—and my praises might go on." [2] Frederick the Great ranked it above his grandest mili-

[1] The original French title of this work appears as "Discours préliminaire des editeurs" at the beginning of the first volume (1751) of the *Encyclopédie, ou Dictionnaire raisonné des sciences, des arts et des métiers, par une société de gens de lettres. Mis en ordre et publié par M. Diderot, . . . et quant à la partie mathématique, par M. d'Alembert, . . .* (Paris, Briasson, David l'aîné, Le Breton, Durand). The seventeen original folio volumes of text of that work were published from 1751 to 1765, the eleven volumes of plates from 1762 to 1772. Four supplementary volumes of text, one of plates, and two of index were published from 1776 to 1780, after Diderot's editorship.

[2] Montesquieu, *Œuvres complètes*, ed. A. Masson (Paris, 1955), III, 1480.

tary accomplishments: "Many men have won battles and conquered provinces," he wrote, "but few have written a work as perfect as the preface to the *Encyclopedia*."[3] Toward the end of the century Condorcet eloquently summed up the judgment of the work shared by a substantial proportion of his fellow *philosophes:* "The union of a vast extent of knowledge, that manner of viewing the sciences which belongs only to a man of genius, a clear, noble, and energetic style, having all the severity which the subject demands and all the pungency that it permits, have placed the *Preliminary Discourse* of the *Encyclopedia* in the number of invaluable works which two or three at the most in each century are in a position to execute."[4] Ever since the eighteenth century the *Discourse* has

[3] Frederick the Great, *Œuvres* (Berlin, 1854), XXV, 166. Letter of 1780 to d'Alembert.

[4] M. J. A. Condorcet, "Éloge de d'Alembert," in *Œuvres complètes de d'Alembert* (Paris, 1821), I, ix, read at the Académie des Sciences and published in 1784. Condorcet (1743–1794), a fellow mathematician and personal friend of d'Alembert, modeled his *Esquisse d'un tableau historique des progrès de l'esprit humain* in part after the *Preliminary Discourse*. The *Esquisse*, written in 1793 while Condorcet was in hiding from the Terror, was the most impassioned statement of faith in progress of the Enlightenment. Voltaire, although he apparently did not immediately recognize the merit of the *Discourse*, later excessively rated it superior to what he called the "method" of Descartes (*Lettres sur Rabelais* [1767], in *Œuvres*, ed. L. Moland [Paris, 1879], XXVI, 513), and in this he was seconded by the nineteenth-century scholar, F. Picavet, who placed the *Preliminary Discourse* on an equal plane with the *Discourse on Method* (1637) in the introduction to the last French edition of d'Alembert's *Discourse* (Paris, 1894), p. lviii. Among the *anti-philosophes* of the eighteenth century who nonetheless praised certain features of the work was even the bitterly hostile Jesuit Father Berthier, editor of the *Journal de Trévoux*, October, 1751, pp. 2270 ff., and October, 1759, p. 2565. The abbé Sabatier de Castres, who consistently attacked the encyclopedists in his *Trois siècles de la littérature françoise* (Paris, 1774), I, 40, regrets that the magnificent promise of the *Discourse* was not fulfilled in the *Encyclopedia* itself. In his *Philosophie du dix-huitième siècle* (Paris, 1818), I, 109–10, J. F. La Harpe, similarly an enemy of the *philosophes*, although he claimed to be a friend of d'Alembert, put him on a level with Pascal as a scientist and writer, because of the excellence of the *Preliminary Discourse*.

been cited repeatedly as the most representative work of its age.[5]

Indeed, the *Preliminary Discourse* could be regarded as the manifesto of the French Enlightenment, at least in the retrospective view of the historian. To be sure, it was not designed to be a pronouncement heralding or justifying revolutionary political action as were the Declaration of Independence, the Declaration of the Rights of Man, and the *Communist Manifesto,* but it expressed the spirit of an intellectual and emotional revolution going on in the eighteenth century that in one way or another lay in the background of each of these. It breathed a confidence that man, through his own intelligent efforts, could transform the conditions of human life and that the beginning of that revolution could already be seen in the sciences and arts.

Compared with anything that had preceded it, the *Discourse* was unique. We have seen its likes since, but one looks in vain throughout previous history for a declaration of principles that represented, as this one did, the views of a party of men of letters who were convinced that through their combined efforts they could substantially contribute to the progress of humanity. Francis Bacon (1561–1626) dreamed of the co-operation of scholars for the advancement of learning, but he had no grounds for hoping that his schemes might soon revolutionize society. His *New Atlantis* was a distant utopia. Descartes (1596–1650) wrote as a courageous, lone individual, seeking truth through the powers of his isolated intelligence, although he of course understood that the advance of knowledge depended on the mutual efforts of scholars. He spoke for

---

[5] Norman C. Torrey, "*L'Encyclopédie* de Diderot, une grande aventure dans le domaine de l'édition," *Revue d'histoire littéraire de la France,* LI (1951), 317, states: "le 'Discours préliminaire' de d'Alembert reste le meilleur resumé de l'esprit du XVIII° siècle." Ernst Cassirer makes ample use of the *Discourse* and other works of d'Alembert in his *Philosophy of the Enlightenment,* tr. C. A. Koelln and J. P. Pettegrove (Princeton, 1951; Beacon paperback, 1955). The list of recent authorities who have made substantial references to the work would be a long one indeed.

himself in the *Discourse on Method,* and not for a group of men of letters. The scholarly societies of the seventeenth and early eighteenth centuries, while hoping to contribute to material progress, were concerned primarily with the erudite and professional activities of closeted savants and did not dream of transforming the conditions of the world in a fundamental way.

However, by the end of the seventeenth century, the members of the European international republic of letters were developing an awareness that cumulatively they were a force in the world, and this birth of a self-conscious sense of power among the literati proved to be one of the revolutionary events of modern times. For the first time large numbers of people were coming to the bracing conclusion that the progress of humanity could be carried forward indefinitely in this world, and men of letters felt they were the prime movers of that progress. Rather than isolating themselves in their closets, certain scholars and writers conceived of themselves as being very much involved in the affairs of the world and believed that their intellectual activity, if it contributed to the progress of knowledge, inevitably served a social function. A capital feature of the great *Encyclopedia* of Diderot, for which the *Preliminary Discourse* was the introduction, was that it became the principal expression of the solidarity and power of that group of energetic and articulate men of letters in France.

As the focal point of the French Enlightenment, the encyclopedic project marked a critical step in the process by which intellectual forces and an ever more influential portion of the French public combined in demanding reform. We now view the *Encyclopedia* more as a major historical event than as an original contribution to any of the branches of knowledge. Its publication precipitated a prolonged controversy that clarified the positions of various amorphous factions in the intellectual and emotional world of the eighteenth century and squarely confronted them with one another. Those who ranged themselves on the side of the encyclopedists in effect formed a party, or the rudiments of one, which served as the spiritual predecessor for groups of reformers who were to preside in large measure over the transformation of the conditions of life in France and throughout the western world during the next generation. Thus, as the most perfect expression of the principles of the encyclopedists and their sympathizers, the *Preliminary Discourse* can be singled out by the historian as the manifesto of the Enlightenment in a fuller sense, perhaps, than d'Alembert or his colleagues might originally have intended it to be.

THE BACKGROUND OF THE *DISCOURSE*

The *Preliminary Discourse* was the work of a young scholar, in association with other men of letters who were not quite angry but were filled with an iconoclastic gusto and a confident dedication to what they felt were great and progressive ideas. Their engaging characters, their colorful and earthy eccentricities growing from an eighteenth-century atmosphere especially conducive to individuality and originality, give a special appeal to the study of that era. The twentieth-century reader is surprised by the degree of compactness and intimacy of the lively universe that existed in Europe just before the vast mushrooming of population during the past two centuries. A very large percentage of the men of letters knew one another well enough to be great friends or to hate one another. The most able of them in France easily gravitated together in Paris. Thus it was possible that in the second half of the 1740's the possessors of perhaps the four best young minds in the realm knew one another and perhaps on occasion even argued, jested, and pontificated around the same table. These were Diderot, the Genevan Rousseau, Condillac, and d'Alembert. Rousseau recalled episodes of their early association during the years before 1750 in his *Confessions*. Diderot, he wrote, became his intimate friend, with whom he had practically daily contact.

I had also begun to see a good deal of the abbé de Condillac who was, like myself, of no consequence in the literary world, but was destined to become what he is at the present day. I was perhaps the first who recognized his ability and saw his true worth. He seemed likewise to take pleasure in my company, and, while I was shut up in my room in the rue Jean-Saint-Denis, near the Opéra, ... he would sometimes come to eat his supper informally along with me, just the two of us. He was then working on his *Essay Concerning the Origin of Human Knowledge* [1746], his first work. When it was completed the problem was to find a publisher willing to accept it. Paris publishers are always arrogant and harsh toward someone who is just beginning, and metaphysics, which was not very fashionable then, hardly seemed an enticing subject. I spoke to Diderot of Condillac and his work and introduced them to one another. They were made to suit each other, and so they did. Diderot induced Durand [6] the publisher to accept the abbé's manuscript, and for his first book this great metaphysician earned—and that almost as a favor—a hundred *livres,* which he might not have received without me. As we lived in quarters that were widely separated, all three of us met once a week at the Palais-Royal, and we dined together at the hôtel du Panier Fleuri. These little weekly dinners must have been very much to Diderot's liking, for he, who nearly always failed to keep his appointments, never missed one of them. There I drew up the plan of a periodical called *Le Persifleur,* to be written alternately by Diderot and myself. I sketched out the first number, and this brought about my acquaintance with d'Alembert, to whom Diderot had spoken of it. Then unforeseen events got in our way, and this plan went no further.

These two authors had just undertaken the *Dictionnaire Encyclopédique,* which at first was only intended to be a kind of translation of Chambers, somewhat like that of James' *Dictionary of Medicine,* which Diderot had just finished. He wanted me to have some part in this second enterprise, and proposed that I should undertake the musical section of it. I consented, and executed it very hastily

[6] This was the publisher who in 1746 purchased Diderot's *Pensées philosophiques,* a book which was immediately condemned to be torn up and burned by the public executioner as being contrary to religion and morals. Later Durand was an associate publisher of the *Encyclopedia* (see n. 1).

and poorly in the three months he had stipulated to me as well as all the others who were to collaborate in the work. But I was the only one who was ready at the prescribed time. I gave him my manuscript, of which I had had a fair copy made by one of M. de Fancueil's lackeys, named Dupont, who wrote very well, paying him ten *livres* out of my own pocket, for which I have never been reimbursed. Diderot, on the part of the publishers, promised me some remuneration, a remuneration of which he has never spoken to me again—nor have I to him.[7]

At the time of the composition and publication of Diderot's *Prospectus* for the *Encyclopedia* (1750) and of the *Preliminary Discourse* (1751), these men, all in their thirties, were just beginning their careers. They belonged to the second generation of the *philosophes*. The time came when they were lionized and wooed by the crowned heads of Europe and the influential in society as the foremost intellectual attractions of France; but these were perhaps their happier days of relative obscurity.

Jean Le Rond d'Alembert (1717–1783), although he was the youngest of the four, was best known for some time because of his precocious scientific and mathematical genius. By the time he joined the encyclopedic group he had already begun to make the contributions in those fields that were to assure him a permanent and important place in the history of science. His origins were bizarre. He was the natural son of a soldier aristocrat, the chevalier Destouches, and Madame de Tencin, one of the most notorious and fascinating aristocratic women of the century. A renegade nun, she acquired a fortune as mistress to the powerful minister, Cardinal Dubois, and after a successful career of political scheming, she rounded out her life by establishing a salon which attracted the most brilliant writers and philosophers of France. It is reported that d'Alembert was not the first of the inconvenient offspring she abandoned. In any case, he was found shortly after his birth on the steps of the Parisian church of Saint-Jean-Lerond. He was

---

[7] Jean-Jacques Rousseau, *Confessions*, ed. Van Bever (Paris, 1926), II, 164–67. The *Confessions* was published posthumously.

raised by a humble nurse, Madame Rousseau, whom he treated as his mother and with whom he lived until long after he achieved international fame. His father, the chevalier Destouches, was able to keep track of the boy and provided him with sufficient means for his schooling. While d'Alembert was yet a child he showed extraordinary promise which matured into genius. For a number of years the young scholar abandoned himself entirely to his passion for the physico-mathematical sciences and, largely without formal training, he succeeded in mastering these fields. At the age of twenty-six he published his *Treatise on Dynamics* (1743), now considered a landmark in the history of Newtonian mechanics, and he continued to contribute significantly to mathematics, astronomy, and dynamics through the period of his greatest scientific productivity in the 1740's and early 1750's. During most of his life he was in intimate contact with the eminent scientists of his day through his correspondence, and as a member of the most distinguished scientific societies of Europe.

D'Alembert's gay and lively character, which shines forth in the famous smiling pastel of him by La Tour in 1753, his enthusiasm, and his brilliance won him the friendship of Diderot and several other excellent Parisian men of letters in the 1740's. At the same time he charmed his way into the center of the powerful salon of Madame du Deffand, who became his intimate friend and protectress. The publication of the *Preliminary Discourse* brought him out of his obscurity as a poverty-stricken mathematician and launched him, in the public mind, as a *philosophe*. He began to be considered a spokesman for the philosophic party, a role which he adopted with great gusto. Endowed with an inquiring and facile intelligence, he easily assimilated the most exciting ideas of the time. His collected works included essays on a remarkable range of subjects. Along with Diderot and Rousseau, he was fascinated by the musical theories of Rameau, as will be seen in the *Discourse*. His *Élémens de musique* (1752), based on the principles of Rameau, was an influential work throughout Europe, and has caused music historians to rate him as one

TRANSLATOR'S INTRODUCTION xvii

of the leading music critics of his century.[8] A combination of virtuosity, ambition, aggressiveness, and personal charm eventually won him a most honored position in the intellectual community of Europe. Among other rewards, it brought him the lasting friendship of both Voltaire and Frederick the Great, who engaged with him in a rich correspondence which lasted from the 1750's throughout the remainder of their lives. Fairly early in his career he was elected to the most important academies of Europe. By the time he joined Diderot as an encyclopedist he was deeply involved in the active and inventive world of the French Academy of Sciences, and that connection was a valuable asset for the encyclopedic project. His entry into the Académie Française in 1754 marked a major victory for the encyclopedic party. Eventually he became the perpetual secretary of that academy, where he followed the tradition of composing eulogies that had been established by Fontenelle when he was secretary of the Academy of Sciences.

Like Diderot and Rousseau, d'Alembert was an emotional man, and as the years passed he became progressively more irascible, even malevolent on occasion. Straitened circumstances and inclination made him abstemious in his habits, and he never married. However, after 1754 he became the in-

---

[8] A. R. Oliver, *The Encyclopedists as Critics of Music* (New York, 1947), p. 157. The most ambitious study of d'Alembert's thought is Maurice Muller's *Essai sur la philosophie de Jean d'Alembert* (Paris, 1926). The standard biography, which leaves much to be desired, is Joseph Bertrand's *D'Alembert* (Paris, 1889). John N. Pappas, in his *Voltaire and d'Alembert* (Bloomington, Indiana, 1962), examines d'Alembert's career as a *philosophe*, for the most part after he ceased to be an active editor of the *Encyclopedia;* Marta Rezler has discussed an important source of information concerning the life of d'Alembert in *The Voltaire-d'Alembert Correspondence: An Historical and Bibliographical Re-Appraisal,* in *Studies on Voltaire and the Eighteenth Century* (Geneva, 1962), XX, 1–139. While this introduction was in press, the publication of Ronald Grimsley's biography of d'Alembert was announced, a work that will no doubt answer the need for a thorough and scholarly study of d'Alembert's life. The standard edition of the writings of d'Alembert is the *Œuvres complètes de d'Alembert* (5 vols.; Paris, 1821-22), which is the one used for the introduction and notes to this work.

timate friend of an aristocratic lady likewise of illegitimate birth, the famous Julie de Lespinasse, with whom he was ever more closely bound until her death left him desolate in 1776. Their strictly spiritual and intellectual relationship, which was something exceptional among the *philosophes,* filled an important chapter of the later part of his life, and together the two friends made a valuable contribution to the vitality of the reforming salons in the period after 1760.

D'Alembert may well have been introduced into his philosophic career by Diderot (1713–1784), with whom he was more intimately allied than with Condillac or Rousseau. Diderot possessed one of the most fascinating intellects of his time, and was, incidentally, one of its great conversationalists. Brilliant, demonstrative, sentimental, endlessly energetic and productive, he was capable of deluging his listener with a flood of surprising and original reflections on almost anything with which his mind grappled. Those around him found themselves hypnotized by his enthusiasm. The son of a cutler from Langres, he had come to Paris as a student and turned to living by his wits as a copyist, translator, and writer. From the outset, it would seem, his tastes were encyclopedic. Among his early productions was a pornographic book, *Les Bijoux indiscrets* (1748), which, judging by the number of editions it has gone through, has turned out to be his most popular work even to this century; but he was also beginning his career as a "philosophical writer" on many subjects that were dangerous to approach in France in the mid-eighteenth century. During his harried life he dashed off works in an astounding variety of subjects and genres: translations, novels, plays, philosophy, scientific theory, psychology, music and drama criticism, social criticism, and polemic. To the chagrin of his friends and the publishers, he was imprisoned in 1749 for some of his unorthodox writings at a time when he very much needed to be free so that he could carry out his multitude of duties as chief editor of the *Encyclopedia.*[9] D'Alembert, by his own account, was

[9] A. M. Wilson's *Diderot: The Testing Years, 1713–1759* (New York, 1957) has the most complete account of the activities of Diderot and his

preoccupied exclusively with mathematics and physics until about the time he came to know Diderot. Then in the middle 1740's he began to move in the circle of the *philosophes* and to convert himself into one of their number, doubtless carried along by the enthusiasms of the men around Diderot.

Jean-Jacques Rousseau (1712–1778) was one of the most extraordinary members of this group. He was a runaway vagabond from Geneva who in 1750 was on the point of becoming a great celebrity in the intellectual world. Passionately fond of music and, like d'Alembert and Diderot, a partisan of the theories of Rameau, he had become a composer himself, as well as the inventor of a new system of musical notation which he had tried to promote in Paris. He and d'Alembert collaborated on the musical articles for the *Encyclopedia*, although Rousseau wrote the bulk of them. He eked out a bare existence as a tutor. Wherever he went he attracted a lively group of companions, but his highly excitable and sensitive soul bore the seeds of a mental disorder which later resulted in a tormented and pathological rupture with many of his old friends, including Diderot and d'Alembert. Rousseau ultimately had a larger personal effect than any of his acquaintances upon the intellectual, emotional, and revolutionary history of Europe, and his career offers one of the best examples of the new impact of the intellectual classes upon the course of history. His democratic political theory became gospel in the political and social revolutions that lay ahead, and his approach to morality, sentiment, and emotion in his novels and other writings lies at the foundation of nineteenth-century Romanticism. However, in 1750 he had only begun to win his first public acclaim in the areas of ethical speculation for which he became famous. In that year he published his *Discourse on the Sciences and the Arts*, which was a subject of comment in the *Preliminary Discourse* (see below, p. 103). How much Rousseau's interest in social speculation helped turn d'Alembert's mind to the same subjects at the time he was writing the *Preliminary Discourse*

---

friends at this period of his life. It has also the best treatment of the history of the *Encyclopedia* to 1759.

we do not know. We may conjecture that all the men in that group of literary acquaintances earnestly discussed such matters and thus had their direct and indirect effects upon one another's thought.

There is no doubt of the effect of the ideas of the abbé de Condillac (1715–1780) upon d'Alembert in the early stages of his intellectual career and later. Condillac was one of a company of men in Paris and elsewhere in France who held the clerical title "abbé." In most cases the title merely signified an intention to become a priest, and it was conferred in a simple ceremony, often when the person who received it was still a child. A substantial share of those who had the title never took holy orders, which would have made them priests. With no real churchly functions, a crowd of abbés converged upon Paris, where many of them graced the households of the wealthy as tutors and witty companions. Some became noted scholars, and Condillac and his brother Mably, who was also an abbé, turned out to be among the most distinguished of them. D'Alembert, being unmarried and poor, seemed particularly to enjoy the company of these intellectual abbés, and he was associated with a number of them who were scholars at the Sorbonne. Although we do not know how intimately he was acquainted with Condillac, we know that he was greatly impressed with the abbé's ideas, and there is an indication that Condillac was influenced by d'Alembert's thoughts on scientific method. Condillac was early fascinated with the analytical method of Locke's psychology, which struck him as the only truly scientific approach to the study of ideas, and he determined to carry it to its logical conclusions. It was in this branch of philosophy that he achieved his greatest eminence. The discussion of the implications of Newton's methods for all philosophical endeavor also formed the object of an important part of his work. He was perhaps the best of the professional philosophers of eighteenth-century France. Condillac's effect upon the thought of the Enlightenment was very great, and it was particularly noticeable in the *Preliminary Discourse*. However, possibly because he did not wish to be compromised by its unorthodoxy, he did not contribute actively to the *Encyclopedia*.

Spreading out beyond this rather bohemian circle of which d'Alembert was a part were the great world of Paris, the rest of France of the Old Régime, and all of Europe, in various states of progress and obscurantism. Scenes of wretchedness, filth, disease and death were daily and public experiences for them all, along with great extremes of luxury and privilege. D'Alembert and his companions lived in a society that was legally defined as one of unequals, in which there existed a host of accretions from the past that had outlasted their usefulness. The general prosperity of almost all classes in France had been rising over the last hundred years. There was a thriving community of professional and business people, who, along with the more intelligent aristocrats and clergy, were responsive to the exchange of ideas and information in literature and to the enjoyments of art. That prosperity made it possible for a man like Diderot to scrape together enough to exist on, albeit miserably, as a professional literary man, and it made it possible in the long run for his ideas and projects and those of men like him to have some real effect in the world which was beginning to be able to afford them.

This is no place to try to sketch the elaborate institutions and abuses of the western Europe of the Old Régime, the old philosophies and interpretations of man and society, and the prejudices and intolerance—or to balance them with a description of the real charms and virtues of that civilization. It must suffice to note that certain forces were at work in all western Europe that progressively brought about fundamental transformations in almost all the important features of that society in the next several generations. France was particularly stirred by the developing spirit of criticism. Through his own study and through his intellectual associations in the Academy of Sciences, the salons, and the circle of Diderot, d'Alembert became caught up in the wave of social and intellectual energy that had agitated the monarchy since the end of the seventeenth century and that was being formed into the philosophy of the Enlightenment by the *philosophes*. The *Preliminary Discourse* itself is one of the best contemporary intellectual records of the enthusiasms shared by d'Alembert and his fellow men of letters during those exciting years when the first volume of the *Encyclopedia* was in preparation. Much of the invigorating thought of the liberal centers of the north had found its way into France, partly through the journalistic activities and translations of the exiled Huguenot intelligentsia, partly through the work of Dutch scientists and critics, and partly through direct communication among various men throughout the international republic of letters. The "English philosophy" of Bacon, Newton, and Locke had become the rage among the *philosophes*. Indeed, d'Alembert was at the forefront of those who truly understood the scientific implications of Newton's work and who transmitted the best of the Newtonian experimental method from England and Holland to France. In addition, he rapidly assimilated a host of stimulating philosophical and social ideas from England, which animated the conversations and writings of his colleagues toward the middle of the century and formed the basis of the *Preliminary Discourse*.

At the same time, d'Alembert and his friends could not have been insensitive to the importance of the intellectual community and tradition of France itself. The many debts of the encyclopedists to their predecessors, and especially to Descartes, are eloquently recognized in the *Preliminary Discourse*, which testified also to the inspiration they drew from the great patriarchs of the Enlightenment: Fontenelle, Montesquieu, and Voltaire. Fontenelle (1657–1757) was still alive in 1750. D'Alembert knew him as a colleague in the Academy of Sciences and sat at his feet in the salon of Madame Geoffrin. All the young *philosophes* were attentive students of the precedents established by Fontenelle for conveying philosophical, critical, and scientific information to the literate public in a brilliant style, and they learned from him how to apply the new methods of philosophy and the discoveries of science to the criticism of various aspects of the social and intellectual heritage of the Old Régime. Similarly, most of them were partisans and disciples of Montesquieu (1689–1755), who had emerged at this time as the great French critic and commentator on the social, political, legal, and constitutional nature of society, and who threw his ponderous influence upon the side of reform. In matters having to do with political theory Montesquieu became the supreme authority for the *Encyclopedia*. Above all, the young *philosophes* admired Voltaire (1694–1778), whose every work of literature, satire, and social criticism stimulated a wave of excitement among the bohemian literati of Paris, and who showed them how to be mordant and "philosophical" when they dealt with the shortcomings and prejudices of their society. The extraordinary eminence of Voltaire was the prime demonstration to the encyclopedists of the power that lay in the hands of men of letters.

Near the middle of the eighteenth century the long development of ideas and attitudes described above reached a critical point in its relationship to the circumstances of French society in general. Widespread disgruntlement was evident against the abuses of the sorry reign of Louis XV; there was a sharpened edge of dissatisfaction in the air. The change that came over certain features of French public opinion was later described in the following terms by a historian who had experienced it: "While we passed from the love of belles-lettres to the love of philosophy, the nation . . . passed over from acclamations to complaints, from songs of triumph to the clamor of perpetual remonstrances, . . . and from a respectful silence regarding religion to importunate and deplorable quarrels. . . . Then it was that there arose among us what we have come to call the *empire of public opinion*. Men of letters immediately had the ambition to be its organs, and almost its arbiters. A more serious purpose diffused itself in intellectual works: the desire to instruct manifested itself in them more than the desire to please. The *dignity of men of letters*, a novel but accurate expression, quickly became an approved expression and one in common use." [10]

The point of view of the intellectuals themselves at mid-

[10] Rulhière's speech to the French Academy, 1787, is reported by A. M. Wilson, *Diderot*, pp. 94–95.

century is best expressed by d'Alembert in his *Élémens de philosophie* of 1759: "If one looks with an attentive eye upon the middle of our century and considers the events which perturb, or at least concern, us, our customs, our works, and even our conversations, it is difficult not to perceive that in several respects a most remarkable change has taken place in our ideas, a change which, by its rapidity, seems to promise us a greater one yet. . . . Our century has called itself . . . supremely the *century of philosophy*. . . . The invention and use of a new method of philosophizing, the kind of enthusiasm which accompanies discoveries, a certain elevation of ideas which the spectacle of the universe produces in us—all these causes must have excited a lively ferment of minds. With a sort of violence this ferment . . . has turned upon everything that was exposed to it like a river which has broken its dikes. . . . Thus, from the principles of the profane sciences to the foundations of revelation, from metaphysics to matters of taste, from music to ethics, from scholastic disputes of theologians to matters of commerce, from the rights of princes to those of peoples, from natural law to the arbitrary laws of nations—in a word, from the questions that touch us most to those which interest us only mildly, everything has been discussed, analyzed, or at least stirred about." [11]

A remarkable cluster of critical works appeared at nearly the precise middle of the century: Montesquieu's *Esprit des Lois* (1748), Buffon's "Premier discours" to his *Histoire naturelle* (1749), Condillac's *Traité des systèmes* (1749), Turgot's *Discours* at the Sorbonne on the progress of the human mind (December, 1750), Rousseau's *Discours [sur les Sciences et les Arts]* (1750), and Voltaire's *Siècle de Louis XIV* (1751). At this conjuncture of landmarks came the first volume of the *Encyclopedia* in June of 1751 with its *Preliminary Discourse* by d'Alembert. Immediately the encyclopedists were recognized as the chief spokesmen of the *philosophes*. We turn now to a brief history of their enterprise.

[11] *Œuvres*, I, 122-23.

At its beginnings the *Encyclopedia* was no more than a venture for profit, projected by a shrewd French publisher. It was to be a translation of the *Cyclopædia* of Ephraim Chambers (1728), an extremely successful English reference work. Only after a hectic and amusing sequence of frauds, quarrels, and blunders were d'Alembert and then Diderot brought into the project, at first as subordinates and finally as editors. Once they were given full control of it, they proceeded to expand its scope vastly and to breathe new inspiration and energy into it. The work was to contain nothing less than the basic facts and the basic principles of all knowledge; it was to be a sort of war machine of the thought and opinion of the Enlightenment.[12]

Neither d'Alembert nor Diderot anticipated that their project would become the center around which the major contending intellectual and emotional factions of France would crystallize. The encyclopedists enjoyed being daring and even shocking, but they were hardly active revolutionaries; one might even say they were rather happy in the world they lived in. Nonetheless, the publication of the *Encyclopedia* became the *cause célèbre* in the intellectual world of the century, almost as the Dreyfus case was in the political world one hundred and fifty years later. Even before the publication of the first volume, an initial sign of trouble appeared in the influential Jesuit periodical, the *Journal de Trévoux,* a few weeks after Diderot's *Prospectus* announced the work, and some time before d'Alembert's *Discourse.* The Jesuit journal published certain criticisms of the claims that were made by Diderot for the projected *Encyclopedia,* and it implied that the encyclopedists had plagiarized from Bacon.[13] Diderot and d'Alembert

---

[12] The history of the *Encyclopedia* is well recorded by Joseph Le Gras in *Diderot et l'Encyclopédie* (Amiens, 1928); F. Venturi, *Le Origini dell' Enciclopedia* (Rome, 1946); A. M. Wilson, *Diderot;* and J. Proust, *Diderot et l'Encyclopédie* (Paris, 1963).

[13] *Journal de Trévoux,* January, 1751, pp. 188–89. The *Prospectus* appeared in November of 1750, announcing the enlarged project of the *Encyclopedia.*

responded angrily. The opening passages of the *Preliminary Discourse* are intended as a refutation of these criticisms.

But the Jesuits had good reason to suspect the orthodoxy of Diderot's religious opinions from the evidence of a number of his works already published, and they had equally good reason to wonder about the attitude of the other contributors to the *Encyclopedia* in matters of religion. They had not themselves been asked to contribute the theological articles; yet they considered themselves the guardians of orthodox theology and defenders of the faith in all respects. They spoke for a very substantial segment of the society of the Old Régime that was growing increasingly concerned about the new "philosophy" and its implications for the doctrines and authority of the Church. Diderot was in an aggressive mood and so were a number of other men of letters, who for a variety of reasons were determined to make a stand for the freedom to express their ideas as they saw fit. The sharp exchange that resulted from the *Prospectus* was one of the first open confrontations between the *philosophes* and the Jesuits.[14] But the Jesuits were only a fraction of the orthodox, although the most alert. The quarrel with them was the opening shot of a long and bitter disputation which broadened out to include larger segments of French society.

The serious opposition to the publication of the *Encyclopedia* came in waves. Churchmen and pious laymen who were alarmed by the ideas that appeared in the first two volumes of the work began to exert pressure on the royal government to put into effect certain long-established measures of repression against dangers to religion, morality, and the authority of the state. At the same time the work was welcomed with widespread enthusiasm, and a host of men of letters and public-spirited amateurs volunteered their services to the enterprise. The scandal caused by the thesis submitted to the Faculty of Theology of the Sorbonne by one of the contributors to the *Encyclopedia,* the abbé de Prades, brought the first

[14] John N. Pappas, *Berthier's Journal de Trévoux and the Philosophes* (Geneva, 1957), pp. 171 ff.; and Wilson, *Diderot,* pp. 117 ff.

major crisis to the enterprise. The thesis, which included whole paragraphs taken directly from d'Alembert's *Preliminary Discourse,* was first accepted and then condemned as heretical.[15] In the atmosphere of shock and repressiveness that resulted from the knowledge that the new philosophy had penetrated even to the citadel of orthodox theology, the first two volumes of the *Encyclopedia* were suppressed, largely because of the suspicious personal and intellectual association between the condemned abbé de Prades and the encyclopedists. Favorable public opinion and a sympathetic faction within the royal government allowed the project to continue once more, but in 1759, after a growing storm of opposition and polemic, the government officially condemned the *Encyclopedia,* and the work could be continued only clandestinely. Many of the contributors, either fearful or exasperated with the persecutions, abandoned the project. D'Alembert refused to have anything more to do with it, except for the strictly mathematical sections. Diderot, with a faithful core of collaborators, of whom the chevalier de Jaucourt (1704–1780) was the chief, secretly carried through work on the last ten volumes of text until their completion in 1765. What had seemed a victory for the enemies of the *philosophes* could not in fact be made decisive, while the strength of those elements of the society sympathetic to the encyclopedists became more and more an accepted fact. By 1765, in an inconclusive way to be sure, the republic of letters associated with the encyclopedists had accomplished a change in the atmosphere of the society of the Old Régime that represented a considerable victory for the new opinions and the power to express them over the objections of the forces of repression. By virtue of the lethargy of the conservatives and the quickening pace of change everywhere, the attitudes toward the world expressed by d'Alembert in the *Preliminary Discourse* were conquering

---

[15] The *Apologie de Mr. l'abbé de Prades* (Amsterdam, 1753) reproduces the thesis in both Latin and French. Franco Venturi, in *Jeunesse de Diderot* (Paris, 1939), pp. 195 ff., includes quotations from the thesis of de Prades that are word-for-word transcriptions of the *Preliminary Discourse.*

an increasing proportion of the effective forces of society. Let us now turn directly to that document.

The *Preliminary Discourse* first appeared, along with Volume One of the *Encyclopedia*, on June 28, 1751. At the outset of his collaboration in the encyclopedic project d'Alembert apparently considered his task to be connected solely with editing the mathematical part of the work. With the exception of two eulogies to fellow academicians, he had not published anything beside scientific works. Why, one might ask, did he, and not Diderot, write the *Preliminary Discourse?* We cannot reply with confidence to that question, for no evidence has come to light yet about the immediate circumstances of d'Alembert's composition of it. We know he was working on it in January, 1751, from a letter of Diderot to Father Berthier of the *Journal de Trévoux*.[16] However, d'Alembert apparently was not contemplating writing an introduction to the *Encyclopedia* at the time of Diderot's imprisonment when the huge project was already rather far along in its preparation. In September of 1749 he wrote Formey, the secretary of the Berlin Academy: "The detention of M. Diderot has become much less severe; nevertheless it still lasts, and the *Encyclopedia* is suspended. I never intended to have a hand in it except for what has to do with mathematics and physical astronomy. I am in a position to do only that, and besides I do not intend to condemn myself for ten years to the tedium of seven or eight folios." [17] Up to this point in his career d'Alembert had never considered himself a serious writer on subjects other than natural science. In a short autobiographical memoir written later in his life he stated that in his passion for mathematics he completely abandoned the culture of belles-lettres which he had formerly enjoyed. He did not return to them, according to his own account, until several years after his entry into the Academy of Sciences (in 1741) at the age of twenty-three, and around the time when he began to work on the *Encyclopedia*,

[16] Diderot, *Correspondance*, ed. G. Roth (Paris, 1955), I, 106.
[17] Quoted in Wilson, *Diderot*, p. 115.

which would be in 1745.[18] One would hardly have expected him to take over the very critical task of composing the definition, so to speak, of the *Encyclopedia* with such scant experience in writing. There may, however, have been strategic reasons. D'Alembert was far better known than Diderot. He was a distinguished member of the three great scientific societies of Europe: the French Academy of Sciences, Frederick the Great's Academy of Berlin, and the English Royal Society. And further, he had no reputation for heterodox and dangerous philosophy, such as Diderot had acquired already in a number of his writings.

The fact that the *Discourse* was d'Alembert's first serious venture outside of mathematics and natural science makes it a truly impressive accomplishment, although he had shown a penchant for "philosophical" writing in the introductions to his scientific treatises, parts of which he transplanted directly into the *Preliminary Discourse*. Later he described the work as "the quintessence of the mathematical, philosophical, and literary knowledge that the author had acquired over twenty years of studies." [19] It seems most likely that Diderot had some hand in it—if not in its actual composition, at least in the discussions that must have gone on concerning it—and his *Prospectus* in a slightly modified form made up the last part of it. We can also be certain that d'Alembert had at his elbow copies of the works of Condillac published up to that time. Indeed, he may have had Condillac himself at his elbow. The work can be considered the product of the conversations and exchanges among these young men, and ultimately of the opinions and information they derived from a very lively atmosphere of criticism and scholarship stretching back to the seventeenth century.[20]

[18] *Œuvres*, I, 2–3; and Bertrand, *d'Alembert*, p. 34.
[19] *Œuvres*, I, 3.
[20] Charles Palissot, an outspoken enemy of the encyclopedists, attempted to cast doubt upon d'Alembert's authorship of the work. In his *Mémoires pour servir à l'histoire de notre littérature* (Paris, 1803), I, 10, he asserts that d'Alembert's friend, the abbé Canaye, remade the entire work from

THE SUBSTANCE OF THE WORK

After giving over a few lines to refuting certain charges made against the encyclopedic project, d'Alembert launches squarely into the substance of his work. It is divided into four sections: (1) The first part examines the genealogy and filiation of ideas and the genesis of the various sciences and arts that will form the basis of the *Encyclopedia*. D'Alembert calls this operation "the philosophic history of the mind." (2) He follows it by a history of intellectual progress since the Renaissance. (3) After this comes a revised version of Diderot's *Prospectus* of November, 1750, then (4) a list of the contributors to the first volume of the *Encyclopedia*. Following the *Discourse* was "A Detailed Explanation of the System of Human Knowledge," which explains the chart or "tree" of knowledge used in the *Encyclopedia,* a short section called "Observations on the Division of Sciences of Chancellor Bacon," an outline of the "General System of Human Knowledge according to Chancellor Bacon," and a graphic chart of the encyclopedic organization of knowledge. Each of these serves as a supplement to the main work. D'Alembert states in his prefaces to the editions of 1753, 1759, and 1764 that Diderot's *Prospectus,* his chart of human knowledge, and the explanation of that chart were a necessary part of the *Discourse*. No doubt Diderot was also responsible for the observations on Bacon and the outline of Bacon's system of knowledge, since both are marked with an asterisk, Diderot's sign of attribution in the *Encyclopedia*. By far the most important parts of the *Discourse* are the two initial sections to which we now turn, where d'Alembert presents the philosophical foundations of the *Encyclopedia* as he discusses the natural history of the mind and makes his eulogy to its progress since the Renaissance.

---

an inadequate and chaotic manuscript of the author. Stories of this kind are common in literary history, and this one deserves no more credence than the rest.

THE METHOD OF THE "DISCOURSE" AND THE "ENCYCLOPEDIA"

The *Encyclopedia* was a response to the growing demand of the intellectual community of Europe since the seventeenth century for a new *summa* of all the branches of knowledge in the light of the major discoveries that had been made in the past one hundred years—a synthesis based upon secular and naturalistic principles rather than upon a traditional theological teleology. Diderot and d'Alembert claimed only to lay the foundations of the true encyclopedia of knowledge in their work, but they hoped at least to propagate the true method of analysis and discovery by which it could be achieved, and to make a contribution to the process of gathering together the essential facts and principles that had been established, or whose validity had been confirmed, by that method. D'Alembert thought he found in the spirit of rigorous analysis, similar to that which produced the recent triumphs of the natural sciences, the means to bind all the disciplines together into a mutually supporting unity that would not be so rigidly systematic as to impose dogmatic limits on the search for new facts. The method and system he envisaged was designed to organize all our valid information and at the same time to facilitate the discovery of more facts and principles that would be useful to humanity.

In the *Discourse* d'Alembert was not only attempting, then, to establish the principles of order upon which the *Encyclopedia* was to be built, he was expounding what might be called the "discourse on method" of the Enlightenment, which he was convinced would progressively give mankind the power of independently shaping and directing its own destiny. In that process he reviewed the history of the progress made thus far in establishing the new "method of reason" and warned of certain pitfalls and errors to be avoided. Out of his discussion we can distill a particular philosophy of history and a number of assumptions that were characteristic of the Enlightenment.

The method of the *Preliminary Discourse* represents an adjustment of the rationalist spirit of Descartes to the empiricism of Locke and Newton—a fusion of traditions which lies at the foundation of the *Encyclopedia*. We may define "rationalism" as an intellectual orientation (particularly notable in the metaphysical systems of the late seventeenth century) which assumes the existence of certain absolute principles or truths instinctively or clearly felt to be true, through which the chief truths of phenomena can be deduced, judged, or explained. "Empiricism" is an intellectual orientation associated with the names of Bacon and Locke, and based on the assumption that "hard facts" of experience, experimentation, and physical sensations are the essential elements from which our valid ideas are derived and which are the source of all true knowledge. Let us look at what d'Alembert borrowed and what he rejected from each of these traditions and their proponents who preceded him.

Although d'Alembert considered Descartes a great intellectual hero because he dared to found philosophy upon systematic doubt and independent human reasoning rather than authority, neither he nor his colleagues could accept a major distinction Descartes had made between the world of physical phenomena and the mind, whereby the one was alleged to be wholly suspectible to the rules of "scientific" study, while the other involved certain innate notions that defied analysis or demonstration. That distinction only fostered supernaturalism and dogmatism in areas they were striving to open for their critical study. The "innate ideas" of Descartes could be used to exempt certain religious, political, and moral domains from critical scrutiny, as priests and traditionalists had done for centuries by arguing that these ideas at the foundation of religion, politics, and morality were implanted by God or innately inscribed on the mind.

The *philosophes* whose opinions d'Alembert expressed did not wish to see any segment of study free of analysis and empirical confirmation. They postulated a division between mind and body, but what they wanted was nothing less than an "experimental physics of the soul," to quote one of the memorable phrases from the *Discourse*. It was John Locke who provided them with its foundations in his analytical sensationalist psychology, which asserted as its chief principle that nothing is in the mind that is not first in the senses. Everything that man knows can be analyzed into simple ideas occasioned by sensations. The great French student of Locke, the abbé de Condillac, presented to his compatriots a combination of certain features of Cartesian rationalism with the empiricism of Locke and Newton that was accepted as doctrine among the *philosophes*, and much of his thought was incorporated into the *Preliminary Discourse*. In d'Alembert's exposition the clear and distinct sensation replaced the clear, distinct, a priori idea of Descartes as the basis of all truth or certain knowledge. With this foundation he thought the *philosophes* could build a complete and unified system of knowledge based upon hard facts and evidence. The laws of any discipline, he was convinced, could be discovered through the analysis of the sensations which produced our ideas, to the point where one that explained all the rest was found.

It should be said that the influence of Bacon upon the *Discourse* and the *Encyclopedia* was perhaps not as great as the pronouncements of the editors might lead one to believe. However enthusiastically d'Alembert may have praised Bacon as the mentor of modern intellectual times and as a proponent of experimentation, it is clear he did not literally accept the extreme inductive method of the English philosopher, which would practically have ruled out the physico-mathematical sciences. For Bacon's intellectual utilitarianism, for his doctrine that knowledge is power, for his vision of mutual cooperation of all scholars in the progress of knowledge, and for his rejection of scholastic teleology, d'Alembert saluted him as a kindred spirit and a great pioneer. Bacon's system of categorizing the various branches of knowledge was useful, with some revision, for the *Encyclopedia*, and Diderot and d'Alembert were much taken with many features of Baconian thought because they found it harmonized with their own. It does not seem, however, that many of the ideas of the *Discourse* were directly or originally occasioned by the English philosopher. In truth, the work was closer to Descartes. D'Alembert was concerned chiefly with the definition and connection of the principles of the various branches of human knowledge in the spirit of Descartes rather than the Baconian collection of information.

From the Cartesians come the empirically undemonstrable metaphysical assumptions of the simplicity and unity of phenomena that run throughout the *Discourse* and the *Encyclopedia*, representing a continuation of the great rationalist tradition of the seventeenth century into the heart of the so-called "empirical" Enlightenment. Much of the argument of the *Discourse* is built upon the assumption of the simplicity of principles or laws of the various disciplines. It is the same kind of simplicity that d'Alembert the geometer demanded in the essential propositions of his own science, and that, as a *philosophe*, he applied as an assumption to the whole range of knowledge. What can be known, he says in effect, can be reduced to simple terms, and this process of analysis into simplest terms can be extended to other branches of knowledge, if not by the application of mathematics or rigorous logic, at least by a process of arranging the facts within a field so that one finds the single fact that explains them all.

Not only are there simple laws behind each discipline, according to the encyclopedists, but there is also a unity that binds them all together. "The rationalistic postulate of unity dominated the minds of this age," wrote Ernst Cassirer, ". . . The function of unification continues to be recognized as the basic role of reason." [21] No one in the eighteenth century expressed that assumption of the unity of all phenomena and all knowledge better than d'Alembert, the geometer turned *philosophe:* "The universe, if we may be permitted to say so, would only be a single fact and a great truth for whoever knew how to embrace it from a single point of view" (p. 29). The

---

[21] Ernst Cassirer, *Philosophy of the Enlightenment*, p. 23.

Encyclopedia represents one step in the effort to achieve that single point of view.

Closely joined with this Cartesian "geometrical spirit" was the fruitful but ultimately undemonstrable idea that all things in nature constitute a continuous chain of being, and that the understanding of one fact is dependent upon an understanding of its relation to others in the continuity of beings in the universe. That notion, too, which Lovejoy traces to its maturity in the eighteenth century, appears throughout the *Encyclopedia* and receives a classic formulation in the *Discourse*. To discover the unknown links of the chain of being and at the same time make them useful for mankind became the program of knowledge of the *philosophes*.[22]

Although they assume at the outset that there are unity, simplicity, and continuity behind all phenomena, d'Alembert and his colleagues do not fall into the snares of the "spirit of systems" of the seventeenth-century rationalist philosophers whom they so much deplored. Repeating the argument of Condillac's *Traité des systèmes* (1749), d'Alembert attacks the metaphysics of such great thinkers as Leibniz and Descartes, who based their philosophical systems upon certain "self-evident" or a priori principles. He tries to clear away a priori metaphysical speculations as undemonstrable and therefore conducive only to error and intellectual despotism. Moreover, he asserts that we know only individual things, that there are no "general beings," and that our general notions are only abstractions of what we observe in individual objects (pp. 48–49, 50). As we have seen, he put forth as the supreme method of achieving truth a combination of rationalism and empiricism closely paralleling the scientific method of the

[22] Arthur O. Lovejoy, *The Great Chain of Being* (Cambridge, Massachusetts, 1948). D'Alembert's article "Cosmologie" in the *Encyclopedia*, vol. IV, presents a memorable statement of the doctrine of continuity, as does the article "Élémens des sciences," vol. V. Nowhere are d'Alembert's prejudices as a geometer more explicitly clear than when he applies the notion of the continuity or chain of geometrical propositions to the whole of phenomena and thus to all knowledge, as he does in the *Preliminary Discourse*, pp. 28–29.

great Newton, which started with sense evidences, or "facts," as the foundation for the discovery of all laws or principles, and which undertook to analyze each object or problem of the physico-mathematical sciences into its simplest elements and thereby to discover the essential principles behind it.[23] The assumptions of the unity, simplicity, and continuity of phenomena in no way prevent the unprejudiced study of the facts presented by the senses. Such are the essentials of the method d'Alembert suggests for the entire encyclopedic range of knowledge in the *Discourse*.

THE FACULTIES OF THE MIND

D'Alembert divides his treatment of the human understanding according to the three faculties of the mind—Memory, Reason, and Imagination—which are operations of the mind defined according to what it does with the perceptions, sensations, or ideas it receives, reflects upon, or puts together. When making his "metaphysical" or logical discussion of the operations of the mind in an individual left in isolation, d'Alembert states that Memory is the first faculty to be exercised, since it involves only the recalling of sensations and ideas passively received by the mind. Reason comes next, involving the comparing, judging, and combining of sensations and the ideas they occasion; and Imagination comes last because it involves the construction of certain new ideas out of the perceptions stored by Memory and the combinations, comparisons, and judgments of Reason. By means of these faculties of the understanding, human beings, according to d'Alembert, historically developed a certain number of disciplines which he categorizes, for the sake of the arrangement of the *Encyclopedia*, beneath headings parallel to those three faculties. Under History fall all the disciplines which have chiefly to do with memory; under Philosophy those which have chiefly to do with

[23] In the physical sciences this process is aided and confirmed by experimentation. See d'Alembert's "Expérimental," *Encyclopédie*, VI, 34 ff., a most sophisticated discussion of the experimental method and its history.

reason; and under Imagination the various Fine Arts. Of course, he points out, the disciplines in the categories of both History and Fine Arts involve reasoning or Philosophy too, and thus are susceptible to the supreme "method of philosophy," the empirical-analytical method which we have discussed earlier.

For all of the disciplines that make up the range of human understanding, d'Alembert asserts there are various means of judging the validity or excellence of the ideas, creations, or actions that constitute their substance. In the more purely intellectual disciplines such as mathematics, the natural sciences, and history, the standards of judgment are derived from the logical comparison of ideas and sense evidences according to the empirical-analytical method. That method is, of course, the prime means of invention, discovery of laws, and extension of the usefulness of all the disciplines involving the operation of reason upon sense evidence. But in those disciplines involving feeling or sentiment, such as ethics and aesthetics, the standards for judging validity and excellence involve both the analytical operations of the mind and something like an *instinctive* judgment of the sentiment which in ethics is conscience and in aesthetics, taste. It is in his discussions of these disciplines that we find the most ambiguity and uncertainty in d'Alembert's exposition, and even something like a return to innate ideas. Moreover, in these areas he is most tempted to assert the existence of certain empirically or logically undemonstrable rights or laws. Nevertheless, he attempts to apply the operations of the method of reason and empiricism to a maximum extent in determining the origins and true principles of our moral and aesthetic sentiments themselves. But he is aware of the limitations of the "method of philosophy" in the study of the knowledge, intellectual creations, judgments, and behavior that are affected by our feelings or sentiments. Indeed, he specifically objects to excessive attempts to submit matters of sentiment and the heart to analysis.

THE "METHOD OF PHILOSOPHY" AND ETHICS

It is apparent in the *Discourse* that d'Alembert believes the rational analysis of human psychological, social, and historical experience might reveal the basic elements of a natural ethics. Through his sketchy remarks we see the outlines of a thoroughly materialistic ethical system which he was perhaps not willing to spell out in full detail. He derives our first and fundamental moral ideas above all from our reactions to the very material sensation of pain, and he baldly states that the sensible moralist should take the sovereign good in this life to be the exemption from pain. Further, he refuses to consider the question of morality and the origins of ethical ideas of man except as he is part of a society. The historical experiences that occasioned the first human ideas of injustice or right and wrong—that is, the first moral ideas—are derived by d'Alembert from the oppression of the weak by the strong in human society. In a brief but surprisingly Hobbesian passage he suggests the existence of a period in human history when men struggled in violent competition for the advantages that society had to offer, a period when brute force was supreme and fear was the dominant emotion (11-12). Although he did not explicitly say it in the *Discourse,* obviously he considered the law of self-preservation or of resistance to oppression to be the foundation of what he called the fundamental "natural law we find within us," the source of all laws and government, and the principle of their necessity.

It was hardly a step from d'Alembert's description of the historical foundation of ethics to a political and moral philosophy that took the general evaluation of pleasure and pain produced by any act as the sole practical and valid means of forming a moral judgment concerning it: and in fact during the course of the French Enlightenment, Helvétius, who was a friend of the encyclopedists although he himself did not contribute to the *Encyclopedia,* carried this line of thought in ethics and political theory to a highly heterodox set of conclusions. The final suppression of the encyclopedic project resulted in part from the association made by the orthodox between the position of the *Encyclopedia* and that of Helvétius, and, indeed, the first suppression of the work in conjunction with the de Prades scandal very much involved ideas to be found in the passages of the *Preliminary Discourse* we have been discussing, passages which were copied in large part directly into de Prades' thesis.

D'Alembert's assertions that the first ideas of right and wrong were derived historically from unhappy social experiences of man, without reference to the Christian tradition, of course brought an alarmed reaction among the orthodox, although he later pointed out that he was speaking only of the *notion*, not the *essence*, of good and evil. The essence presumably comes from God. He was attacked for restricting his discussion of ethics to man as a social being and not in his relation to God, as if there were no ethics without society or relationships of human beings. His description of man's arrival at the idea of right and wrong historically *before* the idea of God only added to the indignation and suspicion. In short, he early encountered the difficulties, in an orthodox society, of viewing society and the history of our ideas and practices strictly from the standpoint of natural evidence, although he publicly denied that this approach rejected or refuted the divine plane of knowledge which he held to be largely beyond our reasoning and surely beyond the scope of the *Encyclopedia*.[24]

[24] The venerable *Journal des Sçavans*, October, 1751, carried a long extract from the work, at the end of which was a series of warnings against the implications of what d'Alembert had to say concerning religion. The author accuses d'Alembert of consciously slighting matters of faith, and points out (pp. 218–23) some of the consequences of the secular morality formulated in the *Discourse*. In the same month of October, 1751, Father Berthier's *Journal de Trévoux* similarly criticized certain points in the *Discourse*, without failing, however, to praise it in other respects. D'Alembert gave over a part of the introduction to volume III of the *Encyclopedia* (1753), pp. viii ff., to a bitter refutation of them, and he did the

D'Alembert's treatment of personal and political morality cannot always be reconciled with the "method of philosophy," as for instance when he speaks of the right of equality (p. 41), which cannot be proved through logic or demonstrated through any empirical means.[25] Moreover, we are left to wonder how he would have reconciled his doctrine of conscience as a feeling that judges the rightness or wrongness of an action with his sensationalist epistemology; the feelings of right and wrong imprinted on men's hearts appear to be a new version of the proscribed innate ideas. Later he expressed indignation at those who criticized him, as he said, for agreeing with Pascal that there were truths that spoke to the heart as well as to the mind.[26] He does not make clear whether this moral conscience universally resulted from the earliest experiences of human beings in society and was taught almost automatically by one generation to the next, or whether it is an inherent part of human nature in the Rousseauian sense. In any case, he seemed to think its basic, valid judgments could be studied in the cumulative historical experience of mankind, that *philosophes* could turn to history and isolate the essentials of human moral nature, and thus the essentials of moral and political science, through the accumulation and comparison of a vast number of cases of human moral behavior.[27]

---

same in the introductions to the 1753, 1759 and 1764 editions of the *Discourse* (*Œuvres*, I, 14–15). "Be Christian," he wrote indignantly, "but on the condition that you will be Christian enough to avoid lightly accusing your brothers of not being so."

[25] The Benthamites, using the utilitarian standard of the greatest good for the greatest number, were able to dispense with the undemonstrable notions of "natural laws of human beings" and "natural rights," and thus avoided a recurrent difficulty in the thinking of the *philosophes*. It is clear that the French Enlightenment provided the intellectual and psychological foundation of the Benthamite utilitarian school of social reform, as it did for most of the other major reform programs that subsequently transformed the globe.

[26] *Œuvres*, I, 15.

[27] See "Élémens des sciences," *Encyclopédie*, V, 495–96.

THE PHILOSOPHY OF HISTORY AND THE IDEA OF PROGRESS IN THE "DISCOURSE"

D'Alembert's tendency to turn to history to establish the fundamentals of a science of ethics reflects a general tendency in the second part of the eighteenth century to look to history for enlightenment concerning the true nature of man and society. The *Discourse* encompasses two separate ways of approaching our ideas. The one approach finds their origins in the operations of the "isolated mind," as distinct from history and inherited experience; the other studies the origins and progress of ideas and of truth in connection with the development and history of society. D'Alembert tries to fuse the two.

Descartes, Locke, and Condillac were chiefly concerned with a purely metaphysical description of the genealogy of ideas, independent of history. Although the *Discourse* derives much of its substance from these men, it also manifests a vital interest—hardly to be found in the speculations of any one of them—in the effect of historical experience on the development of ideas. Even when d'Alembert is writing what he calls the "metaphysical" account of the origin of knowledge in the first part of his essay, he derives the origins of certain ideas from the historical social experience of primitive man. Such is the case with the ideas of right and wrong and those which establish the earliest arts and sciences. And he sets his entire philosophical exposition of the bases of human knowledge in a loosely historical context. However, there are points in this early part of the *Discourse* where d'Alembert switches back and forth abruptly and confusingly from a metaphysical discussion of the genealogy of ideas in the "isolated mind" to a commentary on certain ideas that were produced as a result of historical development (pp. 11 ff.).

In the second part of the essay he moves squarely into the history of the progress of ideas, and in a sense the history of the progress of the method for achieving truth, since the Renaissance. There, to be sure, we are subjected to an untenable description of the origins of the Renaissance and a total rejection of the Middle Ages; and in addition we are affronted by the crude interpretation of history as no more than a series of moral lessons (p. 35). But one observes also in the *Discourse* the elements of a philosophy of history which regarded truth as something that is developed gradually over the generations, and which therefore placed a premium on cultural history and the study of the relationship of ideas to society.

D'Alembert's exposition of modern intellectual progress follows a sequence of stages that emphasize, one after the other, the three chief faculties of the human intelligence—Memory, Imagination, and Reason—which he discussed in his metaphysical description of the mind. Thus he seems to bridge the gap between his "timeless" metaphysics and his historical description of progress. The earlier stages of progress after the beginning of the Renaissance, according to d'Alembert, involved chiefly the revival and development of disciplines having to do first with Memory and then with Imagination. It is as if the exercise of these two faculties were the prerequisite for the revival of the third, Reason, in the historical process of civilization in modern times. The new method of truth or reason, in d'Alembert's exposition, had been something developed through the progressive and cumulative exercise of human faculties over three centuries. He always favored the creative faculties of Imagination and Reason over that of Memory, and he found in modern history that the first, more "primitive," stages of intellectual progress entailed the disciplines of Memory or erudition, whereas only later did the intellectuals and artists of Europe make notable progress in the creations of the faculties of Imagination and Reason. Thus, by his discussion of the evolving disciplines deriving from the various faculties of the human psychology, he provided a way to connect the nonhistorical or abstract metaphysics of the mind and the history of intellectual progress, the lack of which made it hard for Condillac, for instance, to square his psychology with his historiography.

As we have pointed out, d'Alembert's "philosophical" interest in history is a symptom of the deeper and deeper preoccupation of the Enlightenment with the problems of society and the reform of society. Propagandists of the Enlightenment turned to history to find evidence, examples, and confirmations for their opinions, and to discover some meaning in the process for them. They were not interested exclusively in the isolated mind, separated from society and time. Bayle and Fontenelle had anticipated the turn toward history with their critical studies of historical evidence and their use of history as a means of refuting some prejudice or supporting some theory or opinion, and by mid-century their approach to history had taken root, amplified by an ever more critical attitude toward the contemporary institutions and customs of society. Montesquieu's *Esprit des lois* (1748), that monumental study of the principles behind the vast and complicated facts of past and present political institutions, helped bring a fuller appreciation of the need to understand the forces of history. That work was soon followed by the publication of d'Alembert's *Preliminary Discourse,* which occurred at practically the same moment as the delivery of Turgot's *Discourse* on the progress of the human mind at the Sorbonne. Voltaire's *Age of Louis XIV* followed soon afterward in 1751. All three are landmarks in the history of culture, the development of the idea of progress, and in the secular use of history as "philosophy teaching by example."

D'Alembert undertook to combine Condillac's empirical philosophy with an interpretation of intellectual progress throughout history, and to some extent so did Turgot. The remarkable parallels between their two works will strike anyone who reads them together. It is quite possible that these two men knew one another in 1750 and 1751, and we are led to wonder whether there was some direct relationship between their two historical studies. The authors of both shared similar judgments concerning the origins of ideas in the senses, the significance of Bacon and Descartes, the unity and interrelation of all truths, and the relationship of knowledge to society. Both delivered stirring eulogies to the progress of the mind, and both were filled with optimism that this progress would continue. Along with numerous other *philosophes,* both men were excited above all by the progress of the method of knowledge since the time of Bacon, a progress in which each genius was seen to add his contribution to the perfection of that method of critical analysis which gave an ever more perfect key to truth in all levels of life and society, and therefore the greatest hope for progress. Of the two, d'Alembert's work is much the fuller treatment of the intellectual history of Europe since the Renaissance, but it must be said that Turgot expressed a far more sophisticated understanding of the continuity of historical development. He paid more attention to the ancient history of knowledge, and he was far from rejecting the entire Middle Ages and the period of the supremacy of the Church as being a loss to the intellectual progress of mankind. But he was in essential agreement with d'Alembert, and it is not surprising that he later became a contributor to the *Encyclopedia.*[28]

Condorcet, who at the end of the century made the supreme statement of the eighteenth-century idea of progress, recognized his old friends, d'Alembert and Turgot, as his intellectual masters. It might be added that he too was involved in the *Encyclopedia,* although he had been too young to contribute to the original volumes. But his collaboration in the four-volume *Supplement* to the work is one more indication of the extent to which the *Encyclopedia* became the organ and the fulcrum of all those who were apostles of progress and reform in the Old Régime. Condorcet also carried the reforming spirit represented and expressed by the work directly into the Revolution as one of its chief intellectuals before the Terror.

Thus did the *Encyclopedia* lie in the mainstream of modern

---

[28] Turgot, *Œuvres,* ed. G. Schelle (Paris, 1913), I, 214–35. D'Alembert may have known Turgot at the time he wrote the *Preliminary Discourse.* Turgot was one of the group of abbés at the Sorbonne at the time he delivered his discourse on history, and later both he and d'Alembert regularly attended the salons of Madame du Deffand and Madame Geoffrin.

Western civilization. Its editors did not conceive of it as a finished work, but only a step in the march toward truth through the mutual co-operation of men of letters in all the arts and sciences. It was something to build upon; its ideals needed perhaps centuries to be fulfilled before the true encyclopedia of knowledge could be put together. The editors felt as if they were standing at the brink of a new age, where man would have the power to control his own destinies and the material forces of the world. D'Alembert was conscious that the middle epoch of the eighteenth century would be memorable in the history of mankind. The true method of knowledge had been discovered, a method that built upon itself, a philosophy that had discovered the means to correct itself by critical analysis of ever more evidence. There was no end to the possibility of reducing all branches of human knowledge to the principles that united them and in so doing assuring new power, progress, and happiness to mankind. The *Discourse* was a challenge and a manifesto for the future.

THE SPIRIT OF INTELLECTUAL LIBERTY AND SECULARISM IN THE "DISCOURSE"

The *Discourse* was one of the many noble declarations of faith in intellectual freedom produced in the eighteenth century. "Liberty of action and thought alone is capable of producing great things," d'Alembert wrote, "and liberty requires only enlightenment to preserve it from excess" (p. 62). As we have seen, he demanded nothing less than the freedom of scholars to establish most of the intellectual disciplines upon a "natural" and secular foundation, untrammeled by the dictates of Religion and Authority. Clerical control in the intellectual domain was in his opinion the chief enemy. The example of England's liberties, where both political and religious despotism were held in leash, was throughout the *Encyclopedia* his favorite illustration of how things ought to be. In the second part of the *Discourse* d'Alembert triumphantly recorded some of the early victories of the new ideas over that combination of Aristotelian authority and Christian dogma which purported to explain all phenomena definitively in terms of divine purposes and whose partisans were willing to use force, if necessary, to make good their claims. In d'Alembert's words, Descartes was the heroic rebel who first "... dared ... to show intelligent minds how to throw off the yoke of scholasticism, of opinion, of authority—in a word, of prejudices and barbarism.... He can be thought of as a leader of conspirators who, before anyone else, had the courage to arise against a despotic and arbitrary power and who, in preparing a resounding revolution, laid the foundations of a more just and happier government which he himself was not able to see established" (p. 80). Although d'Alembert raised a cry of victory over the defeat of the intellectual despotism of old, by implication he warned that there were still more battles to be won through the co-operation of men of intelligence against the forces of repression, and it is clear that he believed that victory might reform the condition of humanity.

Thus, everywhere throughout the *Discourse* we see d'Alembert firmly practicing his own doctrine of intellectual independence in the full range of encyclopedic enquiry. Theology, once queen of the sciences, is described by him as but an offshoot of Philosophy, and he makes only a half-hearted gesture toward Revelation: "Some truths to be believed, a small number of precepts to be practiced: such are the essentials to which revealed Religion is reduced" (p. 26). He deals with God merely as an abstraction reasoned naturally from the evidence of the senses, and he makes virtually no mention of salvation, the sacraments, the after-world, the Scriptures, or the importance of religious institutions and tradition. Man, not God, is the primary consideration, and the mind of man is the measure of all things. For the Augustinian interpretation of the history of humanity, which looks for a divine plan and origin of all institutions and all historical process, we have seen him substitute a thoroughly secular one. He finds the origin of society and of human intellectual disciplines in man's material needs, ignoring any discussion of divine decrees, and he restricts his treatment to entirely secular and utilitarian sources of the ideas of inequality, law, and government. It was clear to all who read the work that although he was not openly opposed to the practice of religion, at least he was militantly anticlerical, and his anticlericalism led him, along with others of the Enlightenment, to seem to derogate the importance of the very dogmas upon which the priesthood based its power. This was the spirit that characterized the encyclopedic school as a whole.

THE EQUALITARIAN IMPLICATIONS OF THE "DISCOURSE"

Certain ideas in the *Discourse* were not only corrosive to the practical and moral power of the Church in society—they were potentially corrosive to the system of hierarchy and of social and legal privilege that prevailed in the Old Régime. Neither d'Alembert nor his colleagues (with the exception of Rousseau) believed in anything like political democracy, and he had no intention of making his work a political pamphlet advocating changes in the constitution of his society. Yet it must be said that the *Discourse* foreshadowed to some degree the subsequent equalitarian evolution of Western civilization. D'Alembert speaks of the right of equality which was overthrown by force in past history. In effect he asserts that all men are equal in their sensations, which are the source of everything in the mind. He claims the only legitimate principle of distinction among men is an intellectual one, and he waters down even that assertion with the notion that all men can learn anything if they are taught properly, that is, if objects are presented to their sensations properly. The *Encyclopedia* as a whole was not written for the specialist, but was designed to help educate the general public. The editors welcomed their role as popularizers, propagandists, and educators, and this is perhaps a reason they were considered so dangerous by the Church.

One of the prominent features of the *Discourse,* as well as the *Prospectus* and the *Encyclopedia,* was the recognition their authors made of the importance of the tradesman and the artisan, for the encyclopedists were beginning to be in a position to see that technology might transform the conditions of the world more profoundly than anything dreamed of in earlier history. In this respect, d'Alembert seconds Diderot's conviction that it must not be beneath the dignity of the philosopher to pay attention to such useful knowledge for humanity. That the trades and manual arts should have been prominently included in a monumental work of philosophic collection and synthesis of information by intellectuals at the head of the republic of letters was a dramatic gesture and a departure from tradition that had a notable impact upon the imagination of Europe. D'Alembert's statements in the *Discourse* make a fitting philosophical introduction to the renowned descriptions of the arts and trades and the splendid technological plates of the *Encyclopedia*. But he does not try to draw explicitly democratic conclusions in political theory from his recognition of the practical importance and worth of the artisan.

If one examines the folio columns of the *Encyclopedia* he will find that in many respects it fulfills its definition in the *Preliminary Discourse,* and that in other respects it fails. Its dominant philosophy of history follows the lines laid down in d'Alembert's introduction; its barrage of criticisms against scholasticism, the priesthood, and the imperfections of the society of the Old Régime are at certain points far sharper than anything in the *Discourse*. It represents a manful effort to gather together the basic principles and the basic facts of all the arts and sciences, and many of its articles are of exceptional quality.

As might be expected, d'Alembert himself takes up and elaborates on all the themes of the *Discourse* in his many important articles for the *Encyclopedia*. In July of 1751 he replied to Hénault's praises of the *Discourse* in the following terms:

As for myself, I see many shortcomings in my work; I will be happy if I can rectify them, as I intend, in the articles for which I am responsible in the *Encyclopedia*, where I propose to treat more thoroughly a large number of matters which the limits of the *Preliminary Discourse* allowed me to touch only lightly.[29]

D'Alembert's contribution of articles has a particular consistency and unity compared with that of any other contributor, and, so far as it goes, it fulfills the definition of purposes in his introductory essay. Had all parts of the *Encyclopedia* been carried through with equal skill, the work would have been far more satisfactory than it turned out to be. D'Alembert's contribution was unfortunately truncated after his withdrawal in 1758 from everything but the mathematical and physical sections of the work. However, we can get some notion of how he might have carried through his contribution on a truly encyclopedic level in the *Élémens de philosophie* of 1759.

The categories of knowledge defined in the *Discourse* were followed in the body of the work, but it must be admitted that the encyclopedic arrangement based on Bacon's subdivisions had very little revolutionary effect upon the progress of knowledge. The editors presented it only as a provisional principle of organization, and it gave a certain unity and intellectual quality to the work. It served as a vehicle for advertising the virtues of co-operation among the disciplines and for the entire methodology of the Enlightenment. The system of cross references, about which so much was said, was not carried through consistently or perfectly. Nevertheless, they served as a constant reminder of the interrelationship of knowledge and an occasional device to make a covert jab at some notion or institution that the encyclopedists did not dare attack too openly.

There were many, including Diderot himself, who regretted that the splendid promise of the *Preliminary Discourse* was

[29] D'Alembert, *Œuvres*, V, 13–14.

not carried through in the body of the *Encyclopedia*, which was indeed marred by a multitude of imperfections. Nevertheless, it was one of the great enterprises in the history of ideas of Western civilization, and it would have merited the gratitude of humanity if it had been no more than the occasion for the *Discourse* of d'Alembert.

SELECTED BIBLIOGRAPHY

ALEMBERT, JEAN LE ROND D'. *Mélanges de littérature, d'histoire, et de philosophie.* 2 vols., Berlin [Paris], 1753; 5 vols., Amsterdam, 1759–67.

——. *Œuvres complètes de d'Alembert.* 5 vols. Paris, 1821–22.

BECKER, C. *The Heavenly City of the Eighteenth-Century Philosophers.* New Haven, 1954.

BERTRAND, JOSEPH. *D'Alembert.* Paris, 1889.

BRINTON, C. *Ideas and Men.* New York, 1950.

BUTTS, R. E. "Rationalism in Modern Science, d'Alembert and the *Esprit Simpliste*," *Bucknell Review*, VIII (1959), 127–39.

CASSIRER, ERNST. *The Philosophy of the Enlightenment.* Translated by F. C. A. KOELLN and J. P. PETTEGROVE. Princeton, 1951; Beacon paperback edn., Boston, 1955.

HUARD, GEORGES, ed. *Diderot et l'Encyclopédie: Exposition commémorative.* Paris, 1951.

DIECKMANN, HERBERT. *Le Philosophe, Texts and Interpretation.* St. Louis, 1948.

——. "Diderot's Conception of Genius," *Journal of the History of Ideas*, II (1941), 151–82.

FÖRSTER, MAX. *Beiträge zur Kenntnis der Characters und der Philosophie d'Alemberts.* Hamburg, 1892.

FRANKEL, CHARLES. *The Faith of Reason.* New York, 1948.

GILLISPIE, C. C. *A Diderot Pictorial Encyclopedia of Trades and Industry.* 2 vols. New York, 1959.

GORDON, D. H. and TORREY, N. L. *The Censoring of Diderot's Encyclopédie and the Re-established Text.* New York, 1947.

GRIMSLEY, RONALD. *Jean d'Alembert.* Oxford, 1963.

GROSCLAUDE, PIERRE. *Un Audacieux message: L'Encyclopédie.* Paris, 1951.

HAVENS, G. R. *The Age of Ideas.* New York, 1955.

HAZARD, PAUL. *The European Mind, 1680–1715.* Translated by J. LEWIS MAY. London, 1953.

———. *European Thought in the Eighteenth Century.* London, 1954.

HUBERT, RENÉ. *Les Sciences sociales dans l'Encyclopédie.* Paris, 1923.

LA HARPE, J. F. *La Philosophie du dix-huitième siècle.* Vol. I. Paris, 1818.

LE GRAS, JOSEPH. *Diderot et l'Encyclopédie.* Amiens, 1928.

LOUGH, JOHN. *An Introduction to Eighteenth Century France.* London, 1960.

———. "The *Encyclopédie* in Eighteenth-Century England," *French Studies,* VI (1952), 289–308.

LOVEJOY, A. O. *The Great Chain of Being.* Cambridge, Massachusetts, 1948.

MAY, LOUIS-PHILIPPE. "Histoire et sources de l'*Encyclopédie* d'après le registre de délibérations et de comptes des éditeurs et un mémoire inédit" and "Documents nouveaux sur l'Encyclopédie," *Revue de Synthèse,* XV (1938), 7–110.

MULLER, MAURICE. *Essai sur la philosophie de Jean d'Alembert.* Paris, 1926.

NAVES, RAYMOND. *Voltaire et l'Encyclopédie.* Paris, 1938.

OLIVER, A. R. *The Encyclopedists as Critics of Music.* New York, 1947.

PAPPAS, J. N. *Berthier's Journal de Trévoux and the Philosophes.* (Studies on Voltaire and the Eighteenth Century, Vol. III.) Geneva, 1957.

———. "Rousseau and d'Alembert," *PMLA,* LXXV (1960), 47.

———. *Voltaire and d'Alembert.* Bloomington, 1962.

PROUST, J. *Diderot et l'Encyclopédie.* Paris, 1963.

———. "La Documentation technique de Diderot dans l'*Encyclopédie*," *Revue d'histoire littéraire de la France,* LVII (1957), 335–52.

REZLER, MARTA. *The Voltaire–d'Alembert Correspondence.* (*Studies on Voltaire and the Eighteenth Century,* Vol. XX.) Geneva, 1962.

TORREY, N. C. "L'*Encyclopédie* de Diderot, une grande aventure dans le domaine de l'edition," *Revue d'histoire littéraire de la France,* LI (1951), 306–17.

VARTANIAN, ARAM, *Diderot and Descartes: A Study of Scientific Naturalism in the Enlightenment.* Princeton, 1953.

VENTURI, FRANCO. *Jeunesse de Diderot (de 1713 à 1753).* Paris, 1939.

———. *Le Origini dell' Enciclopedia.* Rome, Florence, and Milan, 1946.

WILSON, ARTHUR M. *Diderot, the Testing Years, 1713–1759.* New York, 1957.

NOTE ON THE TRANSLATION

The present translation is based upon the original edition at the beginning of the first volume of the *Encyclopedia* in 1751. The original punctuation and capitalization have generally been modernized. Only minor changes in the *Discourse* were made by d'Alembert in the editions of his *Mélanges de littérature, d'histoire et de philosophie,* published in 1753, 1759, 1760, 1764, and 1770. Where these changes clarify the text, we have noted them. All notes are those of the translators, except where they are otherwise designated. The division into parts is also editorial.

The *Preliminary Discourse* was translated into English in 1752 under the title *The Plan of the French Encyclopædia or Universal Dictionary of Arts, Sciences, Trades and Manufactures,* London. The translation, which was of little use for the present one, was intended as the beginning of an English project for pirating the entire *Encyclopedia*. For a moment this threat alarmed the French publishers, but the enterprise failed. There was one other project to translate the *Encyclopedia,* which came to naught in the same year.[1] Thereafter, only isolated passages of the *Discourse* have been rendered into English in works on the Enlightenment. It has, however, been translated into German, Spanish, and Italian. It went through a number of French editions and reprintings, as a separate work and as a part of the various editions of d'Alembert's works. The supply of the last one, edited by F. Picavet and published by A. Colin (Paris, 1894), which was reprinted a number of times in this century, has long since been exhausted.

This edition and translation was done under the auspices of the Society for Early Modern Studies of Davis, California. It

[1] Wilson, *Diderot,* p. 151; J. Lough, "Le Rayonnement de l'*Encyclopédie* en Grande-Bretagne," *Cahiers de l'Association internationale des Études françaises,* May, 1952, pp. 68–69.

is dedicated to my parents. I wish to thank Professors Merle Perkins and Arthur Child of Davis, Professors Roger Hahn and Carl Schorske of Berkeley, Professors Crane Brinton of Harvard and Arthur M. Wilson of Dartmouth for reading the manuscript and making helpful suggestions. Professor Sherman Stein, Thomas Lehrer, and Professor Jacques Proust of Montpellier provided information on certain technical points. The translation owes whatever merit it possesses to the collaboration of Professor Walter E. Rex of Berkeley, and his criticisms have been invaluable for all parts of the Introduction and notes. Mrs. Lee Barker, Mrs. Linda Hesse, and Mrs. Florence Fisher assisted in the preparation of the manuscript.

RICHARD N. SCHWAB

PRELIMINARY DISCOURSE
TO THE ENCYCLOPEDIA OF DIDEROT

[PART I]

The Encyclopedia which we are presenting to the public is, as its title declares, the work of a society of men of letters. Were we not of their number, we might venture to affirm that they are all favorably known or worthy of being so.[1] But, without wishing to anticipate a judgment which should be made only by scholars, it is at least incumbent upon us, before all else, to remove the objection that could most easily prejudice the success of such a large undertaking as this. We declare, therefore, that we have not had the temerity to undertake unaided a task so superior to our capabilities, and that our function as editors consists principally in arranging materials which for the most part have been furnished in their entirety by others. We had explicitly made the same declaration in the body of the *Prospectus*,[2] but perhaps we should have put it at the beginning of that document. If we had taken that precaution we would doubtless have replied in advance to a large number of gentlemen—and even to some men of letters—who had unquestionably glanced at our *Prospectus*, as their praises attest, but who, nevertheless, have asked us how two persons could treat all the sciences and all the arts.[3] This being

[1] A list of some fifty men who had contributed in one way or another to the first volume of the *Encyclopedia* makes up the last part of the *Discourse*. With each new volume until the final official suppression of the work in 1759, the number of contributors grew in proportion to the enthusiasm for the project among the members of the republic of letters. Very few of the contributors to the first volume of the *Encyclopedia*, except d'Alembert, had achieved a widespread reputation by 1751, and neither Voltaire nor Montesquieu had yet been recruited to the encyclopedic project.

[2] D'Alembert's note as follows: "This *Prospectus* was published in the month of November, 1750."

[3] Most of that *Prospectus* of November, 1750, which announced the enlarged *Encyclopedia* and invited subscribers, is incorporated into the last part of the *Discourse*. D'Alembert refers to the attacks of the *Journal*

the case, the only way of preventing the reappearance of their objection once and for all is to use the first lines of our work to destroy it, as we are doing here. Our introductory sentences are therefore directed solely to those of our readers who will decide not to go further. To the others we owe a far more detailed description of the execution of the *Encyclopedia*, which they will find later in this Discourse, together with the names of each of our colleagues. However, a description so important in its nature and substance must be preceded by some philosophical reflections.

The work whose first volume we are presenting today [4] has two aims. As an *Encyclopedia*, it is to set forth as well as possible the order and connection of the parts of human knowledge. As a *Reasoned Dictionary of the Sciences, Arts, and Trades*, it is to contain the general principles that form the basis of each science and each art, liberal or mechanical, and the most essential facts that make up the body and substance of each.[5] These two points of view, the one of an *Encyclopedia* and the other of a *Reasoned Dictionary*,[6] will thus constitute the basis for the outline and division of our Preliminary Discourse. We are going to introduce them, deal with them one after another,

---

de Trévoux (see notes 13 and 14 of the Introduction). Diderot had promised Father Berthier that d'Alembert would answer all the questions raised in the *Journal*, and many others. See Diderot, *Correspondance*, ed. G. Roth, I, 106.

[4] Later editions: "The work which we are beginning (and which we wish to finish). . . ."

[5] It is essential to remember throughout this work that the word "science" is by no means restricted to natural science. According to Diderot's definition in the *Encyclopedia*, if the object of a discipline is only *contemplated* from different approaches, the technical collection and disposition of observations relative to that object are called "science." Thus theology, history, and philosophy, as well as physics and mathematics, are sciences. If the object of a discipline is something *executed*, the technical collection and disposition of rules according to which it is executed is called "art." By this definition ethics is an art. See "Art," *Encyclopédie*, I, 713–14.

[6] I.e., "Systematic" Dictionary.

and give an account of the means by which we have tried to satisfy this double object.

If one reflects somewhat upon the connection that discoveries have with one another, it is readily apparent that the sciences and the arts are mutually supporting, and that consequently there is a chain that binds them together. But, if it is often difficult to reduce each particular science or art to a small number of rules or general notions, it is no less difficult to encompass the infinitely varied branches of human knowledge in a truly unified system.[7]

The first step which lies before us in our endeavor is to examine, if we may be permitted to use this term, the genealogy and the filiation of the parts of our knowledge, the causes that brought the various branches of our knowledge into being, and the characteristics that distinguish them. In short, we must go back to the origin and generation of our ideas.[8] Quite aside from the help this examination will give us for the encyclopedic enumeration of the sciences and the arts, it cannot be out of place at the head of a work such as this.

---

[7] The rationalist assumption of the unity of knowledge here makes its appearance at the outset of the *Discourse* and, indeed, is the very foundation of the encyclopedic project.

[8] A comparison of d'Alembert's wording with the following lines in Condillac's *Essai sur l'origine des connoissances humaines* (*Essay Concerning the Origin of Human Knowledge,* 1746) gives one indication among many of d'Alembert's debt to the work of Condillac throughout the *Discourse:*

> Our first object, which we should never lose from sight, is the study of the human mind—not to discover its nature, but to learn to know its operations, to observe how they are combined and how we ought to use them in order to acquire all the intelligence of which we are capable. It is necessary to go back to the origin of our ideas, to work out their generation, to follow them to the limits which nature has prescribed for them, and by these means to establish the extent and limits of our knowledge and renew all of human understanding. [*Œuvres philosophiques de Condillac,* ed. Georges Le Roy (Paris, 1947), I, 4.]

Condillac's essay, in turn, was inspired by Locke's great *Essay Concerning Human Understanding* (1690), translated in 1700 into French under Locke's supervision by the Huguenot refugee, Pierre Coste.

We can divide all our knowledge into direct and reflective knowledge. We receive direct knowledge immediately, without any operation of our will; it is the knowledge which finds all the doors of our souls open, so to speak, and enters without resistance and without effort. The mind acquires reflective knowledge by making use of direct knowledge, unifying and combining it.

All our direct knowledge can be reduced to what we receive through our senses; whence it follows that we owe all our ideas to our sensations. This principle of the first philosophers was for a long time regarded as an axiom by the scholastic philosophers. They respected it merely because it was ancient, and they would have defended "substantial forms" and "occult qualities" with equal vigor.[9] Consequently, during the renaissance of philosophy this true principle received the same treatment as the absurd opinions from which it should have been distinguished: it was proscribed along with them, because nothing is more detrimental to truth, and nothing exposes it more to misinterpretation, than the intermingling or proximity of error. The system of innate ideas, which is attractive in several respects, and the more striking perhaps because it was less familiar, replaced the axiom of the scholastic philosophers [that we owe all our ideas to our sensations]; and after having reigned for a long time, the system of innate

[9] The "substantial forms" and "occult qualities" to which d'Alembert refers are technical concepts which he alleges are derived from ancient philosophy and were incorporated into the medieval scholastic philosophical-theological system. The *philosophes* disagreed contemptuously with the scholastics about what was an adequate explanation of any given phenomenon. One of the ways the scholastics "explained" physical phenomena, for instance, was by postulating the existence of "occult qualities" (unknown causes of manifest effects) hidden within bodies. These scholastic inventions are summarily dealt with as being utterly useless explanations of anything in the article "Qualité (Metaphysiq.)," *Encyclopédie*, XIII, 651, which was probably written by d'Alembert. The term "substantial form" is defined in d'Alembert's article "Forme substantielle," *Encyclopédie*, VII, 176, as a "barbarous term of old scholastic philosophy which was principally used to designate alleged material beings which were nevertheless not matter."

ideas still retains some partisans—so great are the difficulties hindering the return of truth, once prejudice or sophism has routed it from its proper place.[10] Of late, however, it has been almost generally agreed that the ancients were right, nor is this the only matter on which we are beginning to draw closer to them.

Nothing is more indisputable than the existence of our sensations. Thus, in order to prove that they are the principle of all our knowledge, it suffices to show that they can be; for in a well-constructed philosophy, any deduction which is based on facts or recognized truths is preferable to one which is supported only by hypotheses, however ingenious. Why suppose that we have purely intellectual notions at the outset [innate ideas], if all we need do in order to form them is to reflect upon our sensations? The discussion which follows will demonstrate that these notions, in fact, have no other origin.[11]

---

[10] Locke made the classic attack upon innate ideas in his *Essay Concerning Human Understanding* (1690), ed. A. C. Fraser (2 vols.; Oxford, 1894), I, cxvi and 38 ff.:

> [They took men away] from the use of reason and judgement, and put them on believing and taking upon trust, without further examination. ... Nor is it a small power it gives one man over another to have the authority to be the dictator of principles and teacher of unquestionable truths, and to make a man swallow that for an innate principle which may serve to his purpose who teacheth them [I, 116].

In the article "Fini," *Encyclopédie*, VI, 817, d'Alembert ridicules those who claim we must have an innate idea of infinity because we have an innate idea of God, who is infinite. They assert we know God before finite creatures and we know creatures only through the idea we have of God, passing from the infinite to the finite. The absurd notion that we derive our knowledge of the finite from our idea of the infinite d'Alembert feels should alone be enough to upset the system of innate ideas.

[11] D'Alembert reproduces here Locke's refutation of the existence of innate ideas, which appears in the following lines of the *Essay Concerning Human Understanding* (I, 38): "Men barely by the use of their natural faculties, may attain to all the knowledge they have without the help of any innate impressions; ..." If one decides that there are no innate ideas, the operations of the mind become susceptible to the precise and measurable critical analysis of physical sense impressions and the simple ideas they produce.

The fact of our existence is the first thing taught us by our sensations and, indeed, is inseparable from them.[12] From this it follows that our first reflective ideas must be concerned with ourselves, that is to say, must concern that thinking principle which constitutes our nature and which is in no way distinct from ourselves. The second thing taught us by our sensations is the existence of external objects, among which we must include our own bodies, since they are, so to speak, external to us even before we have defined the nature of the thinking principle within us. These innumerable external objects produce a powerful and continued effect upon us which binds us to them so forcefully that, after an instant when our reflective ideas turn our consciousness inward, we are forced outside again by the sensations that besiege us on all sides. They tear us from the solitude that would otherwise be our lot. The multiplicity of these sensations, the consistency that we note in their evidence, the degrees of difference we observe in them, and the involuntary reactions that they cause us to experience —as compared with that voluntary determination we have over our reflective ideas, which is operative only upon our sensations themselves [13]—all of these things produce an irresistible impulse in us to affirm that the objects we relate to these sensations, and which appear to us to be their cause, actually exist. This impulse has been considered by many philosophers to be the work of a Superior Being and the most convincing argument for the existence of these objects. And indeed, since

[12] This is a reduction of the famous proof of Descartes: "I think; therefore I am," to "I have sensations; therefore I am." The Cartesians' clear and distinct ideas are replaced by the empiricists' clear and distinct sensations as the basic common denominator of all thought and truth.

[13] This phrase is obscure: ". . . les affections involontaires qu'elles nous font éprouver, comparées avec la détermination volontaire qui préside à nos idées refléchies, et qui n'opère que sur nos sensations mêmes: . . ." It is clarified in d'Alembert's article "Corps," *Encyclopédie,* IV, 262, where he says proof of the existence of matter (bodies) is based "on the *involuntary* effects which they produce in us compared with our *voluntary* reflections on these same sensations."

there is no connection between each sensation and the object that occasions it, or at least the object to which we relate it, it does not seem that any possible passage from one to the other can be found through reasoning. Only a kind of instinct, surer than reason itself, can compel us to leap so great a gap. This instinct is so strong in us that even if one were to suppose for a moment that it subsisted while all external objects were destroyed, the reconstitution of those objects could not add to its strength.[14] Therefore, let us believe without wavering that in fact our sensations have the cause outside ourselves which we suppose them to have; because the effect which can result from the real existence of that cause could not differ in any way from the effect we experience. Let us not imitate those philosophers of whom Montaigne speaks, who, when asked about the principle of men's actions, were still trying to find out whether there are men.[15] Far from wishing to cast shadows on a truth recognized even by the skeptics when they are not debating, let us leave the trouble of working out its principle to the enlightened metaphysicians. It is for them to determine,

[14] Here d'Alembert wrestles with the problem that dogged philosophers from Descartes on who were concerned with the relation of the mind to the exterior world (matter). How can we, limited in all our knowledge to the evidences of our sensations, know that they represent an existence outside themselves and outside ourselves? Descartes and Malebranche state that it is God who assures the harmony between our ideas and the external world. Berkeley resolves the question by denying the existence of matter and asserting that all is idea. For him it is the will of God that puts order, coherence, objectivity, and necessity in the ideas of experience. D'Alembert falls back upon a pragmatic solution to a problem that is logically insoluble in the terms in which it is posed. His critic in the *Journal des Sçavans*, October, 1751, pp. 219–20, complained that d'Alembert did not explicitly assert that God intervenes on the occasion of the sensations and puts in our minds the direct ideas of things and also that God is responsible for our irresistible impulse to believe there is a causal relation between our sensations and the ideas we have of them. An article by a Huguenot theologian in the Dutch edition of the same *Journal*, November, 1751, pp. 404–31, defends the philosophy of Descartes and the Cartesians at great length against d'Alembert's criticisms in the *Preliminary Discourse*.

[15] Montaigne, *Essais*, ed. E. Faguet (Paris, n.d.), I, 171.

if that be possible, what gradation our soul observes when it takes that first step outside itself, impelled, so to speak, and held back at the same time by a crowd of perceptions,[16] which, on the one hand, draw it toward external objects, and on the other hand (since these perceptions belong properly only to the soul itself) seem to circumscribe it in a narrow space from which they do not permit it to withdraw.

Of all the objects that affect us by their presence, the existence of our body strikes us most vividly because it belongs to us most intimately. But hardly do we become aware of the existence of our body before we become aware of the attention it demands of us in warding off the dangers that surround it. Subject to endless needs, extremely sensitive to the action of external bodies, it would soon be destroyed were it not for the care we take in preserving it. Not that all external bodies give us disagreeable sensations; some seem to compensate us by the pleasure which their action brings to us. But such is the misfortune of the human condition that pain is our most lively sentiment; pleasure affects us less than pain and hardly ever suffices to make up to us for it. In vain did some philosophers assert, while suppressing their groans in the midst of sufferings, that pain was not an evil at all.[17] In vain did others place supreme happiness in sensuality—of which they nevertheless deprived themselves through fear of its consequences. All of them would have known our nature better if they had been content to limit their definition of the sovereign good of the present life to the exemption from pain, and to agree that, without hoping to arrive at this sovereign good, we are al-

[16] "Soul" (*âme*), throughout the *Discourse*, means that part of us that thinks, judges, wills, and experiences "feelings" and emotions. For d'Alembert and his contemporaries the idea of the "soul" includes our intellectual nature (mind) as well as our emotional and moral ones; it is the part of us that perceives.

[17] D'Alembert probably refers here to the Stoics, although the line might be applied equally well to the doctrine of optimism which came under widespread fire by the *philosophes*, the most famous attack being Voltaire's *Candide*. The next line doubtless refers to the Epicureans, or to the corruptors of the Epicurean ethics.

lowed only to approach it more or less, in proportion to our vigilance and the precautions we take. Reflections as natural as these will inevitably strike any man who is left alone and free of the prejudices of education or study. They will be the consequence of the first impression he receives from external objects, and they can be placed among the number of those first exertions of the soul which are deemed most valuable and worthy of observation by those who are truly wise, but which are neglected or rejected by ordinary philosophy, whose first principles are almost always contradicted by them.[18]

The necessity of protecting our own bodies from pain and destruction causes us to examine which among the external objects can be useful or harmful to us, in order to seek out some and shun others. But hardly have we begun to survey these objects when we discover among them a large number of beings who seem entirely similar to ourselves, that is, whose forms are entirely like ours and who seem to have the same perceptions as we do, so far as we can judge at first glance. All this causes us to think that they also have the same needs that we experience and consequently the same interest in satisfying them. Whence we conclude that we should find it advantageous to join with them in finding out what can be beneficial to us and what can be detrimental to us in Nature. The communication of ideas is the principle and support of this union, and necessarily requires the invention of signs—such is the source of the formation of societies, with which must have come the birth of languages.[19]

This interchange which so many powerful motives cause us

[18] See Introduction, pp. xxviii ff., for the utilitarian implications of this paragraph.

[19] At this point we move into a section where d'Alembert, in a way not quite parallel to Locke or Condillac, considers the way ideas and knowledge must have developed historically in the earliest collective experiences of mankind, and we see him combining the "metaphysical" with the historical approach to the origin and progress of our ideas, at the same time dealing with the genealogy of ideas in the "isolated mind" and the development of certain ideas resulting from the historical social experience of human beings.

to form with other men soon increases the scope of our ideas and gives birth to entirely new ones in us, by all appearances far removed from those we would have had by ourselves without such aid. It is for the philosopher to decide whether this reciprocal communication, together with the resemblance we note between our sensations and those of our kind, does not contribute much to strengthen our irresistible inclination to assume the existence of all the objects which strike our attention. In order to keep within my subject, I will note simply that the pleasure and the advantage we find in such an interchange, whether in sharing our ideas with other men or in combining their ideas with ours, ought to lead us to strengthen more and more the bonds of the society thus established and to make it as useful for us as possible. But, since each member of society tries to increase for himself the usefulness that he draws from it and must compete with each of the other members, whose eagerness to do the same is equally strong, all cannot have a like share of the advantages, although all have the same right to them. Such a legitimate right is, therefore, soon transgressed by that barbarian right of inequality called the law of the strongest, which we find so difficult not to abuse, though the practice of it likens us to animals. Thus the strength with which nature endows certain men, and which they should doubtless use only to aid and protect the weak, is on the contrary the origin of the latter's oppression. But the more violent the oppression the less patiently the weak suffer it, because they feel that nothing reasonable could have caused their subjection. Such is the origin of the concept of the unjust, and consequently of moral good and evil, of which so many philosophers have tried to find the principle and which the cry of Nature, resounding in all men, makes understood among even the most savage people. Such is the origin also of that natural law which we find within us, the source of the first laws which men must of necessity have created.[20] Even with-

[20] This basic "natural law" is that which naturally leads men to preserve themselves from pain or death at the hands of those who are stronger than they. Ultimately, it is the natural law of social self-preser-

out the help of these laws, that natural law is sometimes strong enough, if not to abolish oppression, at least to hold it within certain limits. Thus the evil we experience through the vices of our own species produces in us the reflective knowledge of the virtues opposed to these vices, a precious knowledge of which we might perhaps have been deprived if a perfect union and equality had prevailed among men.[21]

By the acquired idea of the just and the unjust (and consequently the idea of the moral nature of actions), we are naturally led to consider what principle it is within us that acts, or, which amounts to the same thing, the substance that wills and conceives. It is not necessary to probe deeply into the nature of our body and the idea we have of it to recognize that it could not be that substance, because the properties we observe in matter have nothing in common with the faculty of willing and thinking. Consequently, this being called *Us* is made up of two principles of a different nature [body and soul], so closely united that we could neither suspend nor alter the correspondence which prevails between the movements of the one and the reactions of the other, and subjects each to the other. This mutual slavery [of soul and body] which is so independent of us, together with the reflections we are impelled to make on the nature of the two principles and on their imperfection, lifts us to the contemplation of an all-

---

vation that is stamped upon everyone, and it involves the necessity of resistance to oppression. Thus, the fundamental natural law of ethics and society is, for d'Alembert, an empirical fact in the history of humanity. The first laws of organized societies were designed to limit oppression; thus they derive from the natural law of self-preservation, protecting the members of society as well as possible from pain and death at the hands of the strong. This seems almost to be a Hobbesian view of the natural state of man and the origins of law, government, and the ideas of right and wrong.

[21] This was an especially controversial passage and was roundly attacked for its impious implications in the *Journal des Sçavans*, October, 1751, p. 220, where the author asks how an isolated man learns of his duty to God and himself if the ideas of morality derive only from social experience.

powerful Intelligence who is the source of what we are and who consequently requires our worship.[22] Our inner conviction alone would suffice to make us recognize the existence of such a being, even if the universal testimonies of other men and of all Nature were not joined with it.

It is therefore evident that the purely intellectual concepts of vice and virtue, the principle and the necessity of laws, the spiritual nature of the soul, the existence of God and of our obligations toward him—in a word, the truths for which we have the most immediate and indispensable need—are the fruits of the first reflective ideas that our sensations occasion.

However interesting these first truths may be for the most noble part of ourselves, soon the body, to which the soul is joined, turns our attention to itself, because of the necessity of providing for its endlessly multiplying needs. The care for its preservation must be directed either toward preventing the evils that threaten it or toward remedying those that have attacked it. We try to satisfy these needs by two means: by our own discoveries and by the investigations of other men, which our social intercourse puts us in a position to enjoy. Whence must have come the birth of agriculture and medicine first, and then all the most absolutely necessary arts. They were at the same time both our most primitive knowledge and the source of all other knowledge, even of that which by its nature seems most remote from them. We must develop this point in further detail.

[22] One can see a Cartesian inspiration in parts of this passage. However, in contrast to Descartes, who goes from mind ("I think, therefore I am"), to God, to body, to the rest of the material world, d'Alembert starts from physical sensations, material facts, and the demands of the body. Then he moves to the historical formation of society and ideas of ethics, thereafter to knowledge of the soul, and finally to God. Again he adds historical considerations and Lockeian sensationalism to the timeless Cartesian metaphysical derivation of the origin of ideas. He defends this sequence later against the complaints of the pious by claiming that experience, history, and reason show that the notions of vices and virtues have preceded knowledge of the true God among the pagans (*Œuvres*, I, 14–15).

Working separately or together and assisting one another in their intellectual endeavors, the first men were perhaps not long in discovering a part of the uses to which they could put material bodies. Being eager above all for useful knowledge, they must at first have put aside all idle speculations, and, rapidly surveying the different beings observable in Nature, combined them materially, so to speak, according to their most striking and palpable properties. After this first combining must have come a more sophisticated one, which, although still related to their needs, was chiefly concerned with a deeper study of some of the less evident properties, with the alteration and decomposition of bodies, and with the use which could be derived from them.[23]

Stimulated as they were by so engrossing an aim as self-preservation, these men of whom we speak and their successors were able, no doubt, to make some progress along the path of knowledge. Nevertheless, it was not long before their experience and observations in this vast universe brought them to obstacles which proved insurmountable to their greatest ef-

[23] D'Alembert's abstract terminology makes it difficult to be certain exactly what he means in this passage. Condorcet gives an indication of what is meant by the "first combining" in his *Esquisse,* or *Sketch for a Historical Picture of the Human Mind* (1794; tr. June Baraclough, New York, 1955), pp. 15–16:

> The first fruits of continuous association are a number of arts, all concerned with the satisfaction of simple needs. They include the making of weapons, cooking and construction of the utensils necessary for cooking, preserving food and providing against those times of the year when fresh supplies are unobtainable. . . .

The more sophisticated combining probably refers to the birth of rudimentary sciences. According to René Hubert, *Les Sciences sociales dans l'Encyclopédie* (Lille, 1923), pp. 318–19, this statement is an anticipation of Auguste Comte's assertion that every individual science is born of some corresponding art or craft and that all the sciences have a utilitarian aim and origin. This interpretation would seem to harmonize with what follows in the next few paragraphs of the *Discourse.* Medicine and agriculture, the most immediately necessary arts, are the original source of the physical sciences. In his article "Expérimental," *Encyclopédie,* VI, 298, d'Alembert says that medicine was from the outset the most essential part of the physical sciences.

forts. It was then that their intellects, which had become accustomed to meditation and were eager to draw profit from it, must have made the uniquely interesting discovery of the properties of physical bodies, a discovery that knows no limits and which served as a valuable expedient in these circumstances. And indeed, if an abundance of pleasurable knowledge could console us for our lack of useful truth, we might say that the study of Nature lavishly serves our pleasures at least, even though it withholds from us the necessities of life. It is, so to speak, a kind of superfluity that compensates, although most imperfectly, for the things we lack. Moreover, in the hierarchy of our needs and of the objects of our passions, pleasure holds one of the highest places, and curiosity is a need for anyone who knows how to think, especially when this restless desire is enlivened with a sort of vexation at not being able to satisfy itself entirely. Thus, we owe much of our purely enjoyable knowledge to the fact that we are unfortunately incapable of acquiring the more necessary kind. Another motive serves to keep us at such work: utility, which, though it may not be the true aim, can at least serve as a pretext. The mere fact that we have occasionally found concrete advantages in certain fragments of knowledge, when they were hitherto unsuspected, authorizes us to regard all investigations begun out of pure curiosity as being potentially useful to us. Such was the origin and the cause of progress of the vast science generally called Physics, or the study of Nature, which includes so many different parts. Agriculture and medicine, which were the principal cause of its birth, are nowadays only branches of it, and although they are the most essential and the earliest branches of all, they have been honored more or less in proportion to the degree to which they have been stifled and overshadowed by the others.

In our study of Nature, which we make partly by necessity and partly for amusement, we note that bodies have a large number of properties.[24] However, in most cases they are so

[24] In the next several paragraphs d'Alembert launches into an exposition of his philosophic and scientific method to which he returns again

closely united in the same subject that, in order to study each of them more thoroughly, we are obliged to consider them separately. Through this operation of our intelligence we soon discover properties which seem to belong to all bodies, such as the faculty of movement or of remaining at rest, and the faculty of communicating movement, which are the sources of the principal changes we observe in Nature. By examining these properties—above all the last one—with the aid of our own senses, we soon discover another property upon which all of these depend: impenetrability, which is to say, that specific force by virtue of which each body excludes all others from the place it occupies, so that when two bodies are put together as closely as possible, they can never occupy a space smaller than the one they filled separately. Impenetrability is the principal property by which we make a distinction between the bodies themselves and the indefinite portions of space in which we conceive them as being placed—at least the evidence of our senses tells us such is the case. Even if they are deceptive on this point, it is an error so metaphysical that our existence and the preservation of our lives have nothing to fear from it, and it continually crops up in our mind almost involuntarily, as part of our ordinary way of thinking. Everything induces us to conceive of space as the place (if not real, at least supposed) occupied by bodies. And indeed, it is by conceiving of sections of that space as being penetrable and im-

---

and again in the *Encyclopedia* and in his other works. It is a classic statement of what Cassirer calls the philosophy of the Enlightenment. First the intellect analyzes the objects of rational study into the properties that constitute them, in order to study each of these properties in isolation; then it puts them together once more. Through the process of treating all the data of the disciplines analytically and synthetically, the mind attempts to discover the simple facts or simple laws behind them, which d'Alembert prefers to call "principles." As he and other *philosophes* were well aware, this method had produced the triumphs of mathematics and physics in the seventeenth and eighteenth centuries, among them d'Alembert's own *Traité de dynamique* (1743) and *Traité de l'équilibre et du mouvement des fluides* (1744). They hoped it would produce similar triumphs in the other disciplines. For a further elaboration, see d'Alembert's excellent article "Analytique," *Encyclopédie*, I, 403-4.

mobile that our idea of movement achieves the greatest clarity it can have for us. We are therefore almost naturally impelled to differentiate, at least mentally, between two sorts of extension, one being impenetrable and the other constituting the place occupied by bodies.[25] And thus, although impenetrability belongs of necessity to our conception of the parts of matter, nevertheless, since it is a relative property (that is, we get an idea of it only by examining two bodies together), we soon accustom ourselves to thinking of it as distinct from extension and to considering the latter separately from it.

Through this new consideration we now see bodies only as shaped and extended parts of space, this being the most general and most abstract point of view from which we can envisage them. For extension in which we did not distinguish shaped parts would be only a distant and obscure vision where everything would elude us because we would be unable to discern anything clearly. Color and shape—properties which are always attached to bodies although they vary for each one—help us in some way to detach bodies from the background of space. Even one of these properties is sufficient in this respect, and for the purpose of considering bodies in their most intellectual form, we prefer shape to color. This may be because shape is the most familiar [26] to us, being known both by sight and touch; or because it is easier to conceive of shape without color in a body than color without shape; or finally it may be

[25] Compare with d'Alembert's "Introduction au *Traité de dynamique*" (1743), *Œuvres*, I, 393–94:

> . . . in order to have a clear idea of movement we cannot avoid the necessity of differentiating, at least mentally, two sorts of extension: one which is regarded as impenetrable and which constitutes what is properly called body; the other which, being viewed simply as extension, without examining whether it is penetrable or not, is the measure of the distance from one body to another, and can serve to determine the movement or lack of movement of body.

Newton led the way in this philosophical discussion of the objects of natural science and mathematics, and Locke made several of these same comments about our ideas in his *Essay Concerning Human Understanding*, bk. II, chs. IV and XIII (Fraser's edition, I, 151 ff. and 218 ff.).

[26] Later editions: "more familiar."

because shape serves to fix the parts of space more easily and in a less vague way.

Hence we are led to ascertain the properties of extension simply as to shape. This is the object of Geometry, which facilitates its task by considering extension limited first by a single dimension, then by two, and finally by the three dimensions constituting the essence of an intelligible body (that is to say, of a portion of space terminated in every direction by intellectual boundaries).

Thus, by a few successive operations and abstractions of our minds we divest matter of almost all its sensible properties, in order to envisage in a sense only its phantom. One should recognize first of all that the discoveries to which this investigation brings us will certainly be very useful in all cases in which there is no need to consider the impenetrability of bodies, as, for example, when it is a question of studying their movement, considering them as shaped, mobile parts of space, distant from one another.

Since our examination of shaped extension presents us with a large number of possible combinations, it is necessary to invent some means of achieving those combinations more easily; and since they consist chiefly in calculating and relating the different parts of which we conceive the geometric bodies to be formed, this investigation soon brings us to Arithmetic or the science of numbers. This [science] is simply the art of finding a short way of expressing a unique relationship [a number] which results from the comparison of several others.[27]

---

[27] The definitions of number, arithmetic, and algebra in this paragraph and the succeeding one are derived from Newton (*Arithmetica universalis*). The meanings of the difficult passages in these two paragraphs are clarified in d'Alembert's article "Arithmétique," *Encyclopédie*, I, 675. D'Alembert uses Newton's definition of number as being nothing more than a relationship (*rapport*), and the meaning of several phrases here becomes clearer if the word "number" is substituted for "relationship." Thus, in the next to the last sentence in this paragraph, for example, two is the unique relationship (number) which results from the comparison of the relationships six and four in the following calculation: $6 - 4 = 2$.

The different ways of comparing these relationships [numbers] give the different rules of Arithmetic [addition, subtraction, etc.].

Moreover, if we reflect upon these rules we almost inevitably perceive certain principles or general properties of the relationships, by means of which we can, expressing these relationships [numbers] in a universal way, discover the different combinations that can be made of them. The result of these combinations reduced to a general form will in fact simply be arithmetical calculations, indicated and represented by the simplest and shortest expression consistent with their generality.[28] The science or the art of thus denoting relationships [numbers] is called Algebra. Thus, although no calculation proper is possible except by numbers, nor any magnitude measurable except by extension (for without space we could not measure time exactly), we arrive through the continual generalization of our ideas at that principal part of mathematics, and of all the natural sciences, called the Science of Magnitudes in general. It is the foundation of all possible discoveries concerning quantity, that is to say, concerning everything that is susceptible to augmentation or diminution.

This science is the farthest outpost to which the contemplation of the properties of matter can lead us, and we would not be able to go further without leaving the material universe altogether. But such is the progress of the mind in its investigations that after having generalized its perceptions to the point where it can no longer break them up further into their constituent elements, it retraces its steps, reconstitutes anew its perceptions themselves, and, little by little and by de-

[28] See "Arithmétique universelle," *Encyclopédie*, I, 675–76. We can indicate numbers by general expressions (in algebra these are alphabetical letters). This is what is meant in the phrase "expressing these relationships [numbers] in a universal way." For instance, the letter *a* can be the universal expression for any number and *b* for any other, etc. We can manipulate and combine these alphabetical general numbers in different ways in order to reduce them to their simplest expression possible. See the definition of algebra in the Chart of Knowledge (p. 145).

grees, produces from them the concrete beings that are the immediate and direct objects of our sensations. These beings, which are immediately relative to our needs, are also those which it is most important for us to study. Mathematical abstractions help us in gaining this knowledge, but they are useful only insofar as we do not limit ourselves to them.

That is why, having so to speak exhausted the properties of shaped extension through geometric speculations, we begin by restoring to it impenetrability, which constitutes physical body and was the last sensible quality of which we had divested it. The restoration of impenetrability brings with it the consideration of the action of bodies on one another, for bodies act only insofar as they are impenetrable. It is thence that the laws of equilibrium and movement, which are the object of Mechanics, are deduced. We extend our investigations even to the movement of bodies animated by unknown driving forces or causes, provided the law whereby these causes act is known or supposed to be known.[29]

Having at last made a complete return to the corporeal world, we soon perceive the use we can make of Geometry and Mechanics for acquiring the most varied and profound knowledge about the properties of bodies. It is approximately in this way that all the so-called physico-mathematical sciences were born. We can put at their head Astronomy, the study of which, next to the study of ourselves, is most worthy of our application because of the magnificent spectacle which it presents to us. Joining observation to calculation and elucidating the one by the other, this science determines with an admirable precision the distances and the most complicated movements of the heavenly bodies; it points out the very forces by which these movements are produced or altered. Thus it may justly be regarded as the most sublime and the most reliable application of Geometry and Mechanics in combina-

[29] For example, the movement of the planets, whose ultimate cause is unknown, although the laws of their movement are known, or at least measurable.

tion, and its progress may be considered the most incontestable monument of the success to which the human mind can rise by its efforts.

The use of mathematical knowledge is no less considerable in the examination of the terrestrial bodies that surround us. All the properties we observe in these bodies have relationships among themselves that are more or less accessible to us. The knowledge or the discovery of these relationships is almost always the only object we are permitted to attain, and consequently the only one we ought to propose for ourselves. Thus, it is not at all by vague and arbitrary hypotheses that we can hope to know nature; it is by thoughtful study of phenomena, by the comparisons we make among them, by the art of reducing, as much as that may be possible, a large number of phenomena to a single one that can be regarded as their principle. Indeed, the more one reduces the number of principles of a science, the more one gives them scope, and since the object of a science is necessarily fixed, the principles applied to that object will be so much the more fertile as they are fewer in number.[30] This reduction which, moreover, makes them easier to understand, constitutes the true "systematic spirit." One must be

---

[30] This sentence is a direct quotation from d'Alembert's "Introduction au *Traité de dynamique*" (1743), *Œuvres*, I, 392. Aram Vartanian, in *Diderot and Descartes* (Princeton, 1953), p. 157, sees this part of the *Discourse* as a continuation of Descartes' rationalistic ideal of relating the totality of physical data to a unique cause. Condillac seems to have incorporated ideas from this same passage of d'Alembert's almost word for word into the first paragraph of his *Traité des systèmes* (1749), *Œuvres*, I, 121:

> A system is nothing more than the arrangement of different parts of an art or a science in an order in which they all support one another, and in which the last are explained by the first. Those which explain the others are called *principles;* and the system is so much the more perfect as the principles are fewer in number: it is even to be hoped that one could reduce them to a single one.

D'Alembert in his turn uses the material from Condillac's work on the "spirit of system" for making some of the major points in his *Discourse*. (See pp. 94–95.) One is led to wonder whether the two men did not collaborate to some extent on both works.

very careful not to mistake this for the "spirit of system," with which it does not always agree. We will speak more fully of this matter later.

But as the reduction of which we speak becomes more or less laborious in proportion to the degree of difficulty and the vastness of the object we embrace, so likewise we will have a greater or lesser right to demand of those who apply themselves to the study of Nature that they furnish us with such a reduction. For example, the magnet, one of the bodies that has been studied the most and about which such surprising discoveries have been made, has the property of attracting iron and of communicating this property to it. It turns toward the poles of the world with a variation which is itself subject to rules and is no less astonishing than a more exact direction would be. And finally, it has the property of forming a greater or lesser angle of inclination with the horizontal line according to the location on the globe where it is placed. All these singular properties, dependent on the nature of the magnet, probably relate to some general property which is the origin of them all and which up to now is unknown to us and perhaps will remain so for a long time. Since such knowledge and the necessary enlightenment concerning the physical cause of the properties of the magnet are lacking, it would doubtless be an investigation most worthy of a philosopher to reduce, if possible, all these properties to a single one, while showing the liaison that they have with one another. But the more useful such a discovery would be to the progress of physical science, the more we have occasion to fear that it will elude our efforts. I say the same of a large number of other phenomena whose enchainment perhaps belongs to the general system of the universe.

The only resource that remains to us in an investigation so difficult, although so necessary and even pleasant, is to collect as many facts as we can, to arrange them in the most natural order, and to relate them to a certain number of principal facts of which the others are only the consequences. If we presume sometimes to raise ourselves higher, let it be with that wise circumspection which befits so feeble an understanding as ours.

Such is the plan we must follow in that vast part of physics called General and Experimental Physics.[31] It differs from the physico-mathematical sciences in that it is properly only a systematic collection of experiments and observations. On the other hand, the physico-mathematical sciences, by applying mathematical calculations to experiment, sometimes deduce from a single and unique observation a large number of inferences that remain close to geometrical truths by virtue of their certitude. Thus, a single experiment on the reflection of light produces all of Catoptrics, or the science of the properties of mirrors. A single experiment on the refraction of light produces the mathematical explanation of the rainbow, the theory of colors, and all of Dioptrics, or the science of [32] concave and convex lenses. From a single experiment on the pressure of fluids, we derive all the laws of their equilibrium and their movement. Finally, a single experiment on the acceleration of falling bodies opens up the laws of their descent down inclined planes and the laws of the movements of pendulums.

It must be confessed, however, that geometers sometimes abuse this application of algebra to physics. Lacking appropriate experiments as a basis for their calculations, they permit themselves to use hypotheses which are most convenient, to be sure, but often very far removed from what really exists in Nature. Some have tried to reduce even the art of curing to calculations, and the human body, that most complicated machine, has been treated by our algebraic doctors as if it were the simplest or the easiest one to reduce to its component parts. It is a curious thing to see these authors solve with the stroke of a pen problems of hydraulics and statics capable of occupying the greatest geometers for a whole lifetime. As for us who are wiser or more timid, let us be content to view most of these calculations and vague suppositions as intellectual

[31] Chemistry would fall under this category of General and Experimental Physics.

[32] Later editions: "the science of the properties of."

games to which Nature is not obliged to conform, and let us conclude that the single true method of philosophizing as physical scientists consists either in the application of mathematical analysis to experiments, or in observation alone, enlightened by the spirit of method, aided sometimes by conjectures when they can furnish some insights, but rigidly dissociated from any arbitrary hypotheses.[33]

Let us stop here a moment and glance over the journey we have just made. We will note two limits within which almost all of the certain knowledge that is accorded to our natural intelligence is concentrated, so to speak.[34] One of those limits, our point of departure, is the idea of ourselves, which leads to that of the Omnipotent Being, and of our principal duties. The other is that part of mathematics whose object is the general properties of bodies, of extension and magnitude. Between these two boundaries is an immense gap where the Supreme Intelligence seems to have tried to tantalize the human curiosity, as much by the innumerable clouds it has spread there as by the rays of light that seem to break out at intervals to attract us. One can compare the universe to certain works of a sublime obscurity whose authors occasionally bend down within reach of their reader, seeking to persuade him that he understands nearly all. We are indeed fortunate if we do not lose the true route when we enter this labyrinth! Otherwise the flashes of light which should direct us along the way would often serve only to lead us further from it.)

The limited quantity of certain knowledge upon which we can rely, relegated (if one can express oneself this way) to the two extremities of space to which we refer, is far indeed from being sufficient to satisfy all our needs. The nature of man, the

[33] Throughout this discussion we have a restatement of several points of Condillac's *Traité des systèmes*, in *Œuvres*, I, 199 ff.

[34] Throughout the *Discourse*, d'Alembert is reflecting a major concern of Bacon, Descartes and Locke, who had all been occupied with the question of establishing the limits of our knowledge so that philosophers might most profitably expend their efforts on what can be known through reason, rather than wasting them, in flights of speculation, imagination, and superstition, on what cannot.

study of which is so necessary and so highly recommended by Socrates, is an impenetrable mystery for man himself when he is enlightened by reason alone; and the greatest geniuses, after considerable reflection upon this most important matter, too often succeed merely in knowing a little less about it than the rest of men. The same may be said of our existence, present and future, of the essence of the Being to whom we owe it, and of the kind of worship he requires of us.

Thus, nothing is more necessary than a revealed Religion, which may instruct us concerning so many diverse objects. Designed to serve as a supplement to natural knowledge, it shows us part of what was hidden, but it restricts itself to the things which are absolutely necessary for us to know. The rest is closed for us and apparently will be forever. A few truths to be believed, a small number of precepts to be practiced: such are the essentials to which revealed Religion is reduced. Nevertheless, thanks to the enlightenment it has communicated to the world, the common people themselves are more solidly grounded and confident on a large number of questions of interest than the sects [35] of the philosophers have been.

With respect to the mathematical sciences, which constitute the second of the limits of which we have spoken, their nature and their number should not overawe us. It is principally to the simplicity of their object that they owe their certitude. Indeed, one must confess that, since all the parts of mathematics do not have an equally simple aim, so also certainty, which is founded, properly speaking, on necessarily true and self-evident principles, does not belong equally or in the same way to all these parts. Several among them, supported by physical principles (that is, by truths of experience or by simple hypotheses), have, in a manner of speaking, only a certitude of experience or even pure supposition. To be specific, only those that deal with the calculation of magnitudes and with the general properties of extension, that is, Algebra, Geometry, and Mechanics, can be regarded as stamped by the seal of evidence. Indeed, there is a sort of gradation and shad-

---

[35] Later editions: "all the sects."

ing, so to speak, to be observed in the enlightenment which these sciences bestow upon our minds. The broader the object they embrace and the more it is considered in a general and abstract manner, the more also their principles are exempt from obscurities. It is for this reason that Geometry is simpler than Mechanics, and both are less simple than Algebra. This will not be a paradox at all for those who have studied these sciences philosophically. The most abstract notions, those that the common run of men regard as the most inaccessible, are often the ones which bring with them a greater illumination. Our ideas become increasingly obscure as we examine more and more sensible properties in an object. Impenetrability, added to the idea of extension, seems to offer us an additional mystery; the nature of movement is an enigma for the philosophers; the metaphysical principle of the laws of percussion is no less concealed from them. In a word, the more they delve into their conception of matter and of the properties that represent it, the more this idea becomes obscure and seems to be trying to elude them.[36]

Thus one can hardly avoid admitting that the mind is not satisfied to the same degree by all the parts of mathematical knowledge. Let us go further and examine without bias the essentials to which this knowledge may be reduced. Viewed at first glance, the information of mathematics is very considerable, and even in a way inexhaustible. But when, after having gathered it together, we make a philosophical enumeration of it, we perceive that we are far less rich than we believed ourselves to be. I am not speaking here of the meager application and usage to which a number of these mathematical truths lend themselves; that would perhaps be a rather feeble argument against them. I speak of these truths considered in themselves. What indeed are most of those axioms of which Geometry is so proud, if not the expression of a single simple idea by means

---

[36] This celebrated paragraph comes directly from d'Alembert's "Introduction au *Traité de dynamique*," *Œuvres*, I, 391–93, showing he had a taste for metaphysical and philosophical statements well before he wrote the *Discourse*.

of two different signs or words? Does he who says two and two equals four have more knowledge than the person who would be contented to say two and two equals two and two? Are not the ideas of "all," of "part," of "larger," and of "smaller," strictly speaking, the same simple and individual idea, since we cannot have the one without all the others presenting themselves at the same time? As some philosophers have observed, we owe many errors to the abuse of words. It is perhaps to this same abuse that we owe axioms. My intention is not, however, to condemn their use; I wish only to point out that their true purpose is merely to render simple ideas more familiar to us by usage, and more suitable for the different uses to which we can apply them. I say virtually the same thing of the use of mathematical theorems, although with the appropriate qualifications. Viewed without prejudice, they are reducible to a rather small number of primary truths. If one examines a succession of geometrical propositions, deduced one from the other so that two neighboring propositions are immediately contiguous without any interval between them, it will be observed that they are all only the first proposition which is successively and gradually reshaped, so to speak, as it passes from one consequence to the next, but which, nevertheless, has not really been multiplied by this chain of connections; it has merely received different forms. It is almost as if one were trying to express this proposition by means of a language whose nature was being imperceptibly altered, so that the proposition was successively expressed in different ways representing the different states through which the language had passed. Each of these states would be recognized in the one immediately neighboring it; but in a more remote state we would no longer make it out, although it would still be dependent upon those states which preceded it and designed to transmit the same ideas. Thus, the chain of connection of several geometrical truths can be regarded as more or less different and more or less complicated translations of the same proposition and often of the same hypothesis. These translations are, to be sure, highly advantageous in that they put us in a position to make various uses of the theorem they express—uses more estimable or less, in proportion to their import and consequence. But, while conceding the substantial merit of the mathematical translation of a proposition, we must recognize also that this merit resides originally in the proposition itself. This should make us realize how much we owe to the inventive geniuses who have substantially enriched Geometry and extended its domain by discovering one of these fundamental truths which are the source and, so to speak, the original of a large number of others.

It is the same with the physical truths and with the properties of bodies whose connection we perceive. All of these properties gathered together offer us, properly speaking, only a simple and unique piece of knowledge. If others in larger quantity seem detached to us and form different truths, we owe this sorry advantage to the feebleness of our intelligence, and we may say that our abundance in that regard is the effect of our very poverty. Electrical bodies, in which so many curious but seemingly unrelated properties have been discovered, are perhaps in a sense the least known bodies, because they appear to be more known. That power of attracting small particles which they acquire when they are rubbed, and that of producing a violent commotion in animals, are two things for us. They would be a single one if we could reach the primary cause. The universe, if we may be permitted to say so, would only be one fact and one great truth for whoever knew how to embrace it from a single point of view.[37]

The different parts of knowledge, whether useful or pleasing, of which we have hitherto spoken and which owe their origin to our needs, are not the only ones which must have

[37] In these paragraphs, as d'Alembert moves from a description of the unity of geometry to the statement of the assumption of unity in all phenomena, we have a most perfect exposition of the marriage of the "geometric spirit" with the empiricism for which Condillac gave the theoretical definition in his *Traité des systèmes* (1749). There Condillac asserted that the only legitimate system was one that explained facts by facts, and the best system was one in which one fact would explain all the others.

been cultivated. There are other branches of knowledge which are related to them and to which, therefore, men applied themselves at the same time they applied themselves to the former. We would have spoken of them all at the same time had we not thought it more appropriate and more in harmony with the philosophic order of this Discourse to consider first and without interruption the general study that men have made of bodies, the study by which men began, although other investigations were soon joined to it. Here is the approximate order in which these other studies must have followed one another.

The advantage men found in enlarging the sphere of their ideas, whether by their own efforts or by the aid of their fellows, made them think that it would be useful to reduce to an art the very manner of acquiring information and of reciprocally communicating their own ideas. This art was found and named Logic. It teaches how to arrange ideas in the most natural order, how to link them together in the most direct sequence, how to break up those which include too large a number of simple ideas, how to view ideas in all their facets, and finally how to present them to others in a form that makes them easy to grasp. This is what constitutes this science of reasoning, which is rightly considered the key to all our knowledge. However, it should not be thought that it [the formal discipline of Logic] belongs among the first in the order of discovery. The art of reasoning is a gift which Nature bestows of her own accord upon men of intelligence, and it can be said that the books which treat this subject are hardly useful except to those who can get along without them. People reasoned validly long before Logic, reduced to principles, taught how to recognize false reasonings, and sometimes even how to cloak them in a subtle and deceiving form.[38]

[38] In this rather confusing paragraph, d'Alembert is distinguishing between the faculty for thinking logically, which people have always had, and the formal discipline (Logic) that analyzes and makes explicit its rules. Logic is only one of the several meanings attached to the words "reason" and "reasoning" in the Enlightenment. The term "reason" came

This most valuable art of assigning a suitable connection to ideas, and consequently of facilitating the passage from one to the other, furnishes the means of effecting at least a partial reconciliation among men who seem to differ the most. Indeed, all our knowledge is ultimately reduced to sensations that are approximately the same in all men. And the art of combining and relating direct ideas in reality adds nothing to them except a more or less exact arrangement and an enumeration which can be rendered more or less intelligible to others. The difference between the man who combines ideas easily and the one who combines them with difficulty is scarcely greater than that between the man who judges a picture at one glance and the one who needs to have all the parts pointed out to him one after another in order to appreciate it. Both have experienced the same sensations in their first glance, but these sensations have only slid over the second man, so to speak. To lead him to the same point at which the other man arrived instantaneously, it would only have been necessary to stop him and make him concentrate for a longer time on each sensation. By this means, the reflective ideas of the first man would have become as much within reach of the second as the direct ideas. Hence it is perhaps true that there is hardly a science or an art which cannot, with rigor and good logic, be taught to the most limited mind, because there are but few arts or sciences whose propositions or rules cannot be reduced to some simple notions and arranged in such a close order that their chain of connection will nowhere be interrupted. As the mind operates more or less slowly, that chain will be required in greater or less degree, and the only advantage possessed by great geniuses [39] is that they have less need of it than others, or rather they are able to form it rapidly and almost unconsciously.

---

to apply to the entire method of analysis and synthesis described up to this point in the *Discourse*.

[39] D'Alembert uses at least two different meanings for the word "genius" in this *Discourse:* the older notion of genius as some excellent quality which certain persons had the good fortune to share (found in the phrase: "He has genius"), and the more romantic one beginning to ap-

The science of communication of ideas is not confined to putting order in ideas themselves. In addition it should teach how to express each idea in the clearest way possible, and consequently how to perfect the signs that are designed to convey it; and indeed this is what men have gradually done. Languages, born along with societies, doubtless began as only a rather bizarre collection of signs of all sorts, and consequently, the natural bodies which impinge upon our senses were the first objects to be designated by names. But so far as we can judge, languages in this first beginning [40] were intended only for the most pressing use, and they must have been very imperfect and limited, and subject to very few definite principles. The arts or sciences which were of absolute necessity may have made considerable progress while the rules of diction and of style were yet to be born. The communication of ideas hardly suffered in any case from this lack of rules or even from the paucity of words; or rather it suffered only to the degree required to oblige each man to augment his own knowledge by stubborn effort, without depending too much upon others. Too much communication can sometimes benumb the mind and prejudice the efforts of which it is capable. If one observes the prodigies of some of those born blind, or deaf and mute, one will see what the faculties of the mind can perform if they are lively and called into action by difficulties which must be overcome.

However, since there are some incontestable advantages in being able to convey and receive ideas easily in mutual intercourse, it is not surprising that men have sought more and more to augment that facility. To do so, they began by re-

---

pear in the eighteenth century, which assigns the name "genius" to a *person* of transcendent creativity, intellect, or ability (see pp. 26, 29, 34, 39, 42, 51, 60, 75). Saint-Lambert's celebrated article "Génie," *Encyclopédie*, VII, 582–84, carries through this latter definition in detail. In addition d'Alembert defines genius as a "feeling that creates" (45). For a lucid discussion of the concept of genius in eighteenth-century France, see Herbert Dieckmann, "Diderot's Conception of Genius," *Journal of the History of Ideas,* II (1941), 151–82.

[40] Later editions: "formation."

ducing signs to words, because words are, so to speak, the symbols that we have most readily at hand. In addition, the order in which words are generated followed the order of the operations of the mind: after individual objects, names were given to the sensible qualities, which, though they do not exist by themselves, exist in these individual objects and are held in common by several of them. Little by little those abstract terms were achieved, some of which serve to tie ideas together, others to designate the general properties of bodies, and others to express purely intellectual notions. All those terms which children take such a long time to learn undoubtedly took even longer to find. Finally, men reduced the usage of words to precepts and formed Grammar, which can be regarded as one of the branches of Logic. When enlightened by a subtle and refined Metaphysics, Grammar separates the shadings of ideas and teaches how to distinguish these shadings through different signs. It sets rules so that signs can be utilized to the greatest advantage, and often discovers, through that philosophic spirit which delves back to the source of everything, the reasons for the apparently bizarre selection which makes one sign preferred to another. And finally it leaves to that national caprice called "usage" only what absolutely cannot be taken away from it.[41]

While communicating ideas to one another, men try also to communicate their passions. They succeed in doing this by Eloquence. As Logic and Grammar speak to the mind, Eloquence was created to speak to sentiment, and can impose

[41] D'Alembert reproduces here some of the interest in the development of language which Condillac found to be of central importance in his discussion of ideas (Part Two of his *Essai sur l'origine des connoissances humaines* [1746], in *Œuvres*, I, 60 ff., which in turn owes something to Locke's discussion of signs and Warburton's discussions of the origins of language). The history of the development of signs is, in effect, the history of ideas, starting in earliest times with words for individual things, then moving to abstract words involving general properties of objects. These also indicate the birth of first the arts and then the sciences. D'Alembert carries on the discussion of signs in the *Encyclopedia* (cf. "Caractere," vol. II).

silence even upon reason. The prodigious effect that it has worked upon an entire nation, often through a single individual, is perhaps the most striking evidence of the superiority of one man over another. It is most surprising that men have believed they could make rules take the place of such a rare talent. It is almost as if they wanted to reduce genius to precepts. The man who first alleged that we owe orators to art either was not one of their number or was quite ungrateful to Nature. Nature alone can create an eloquent man. Men themselves are the first book that he must study in order to succeed; the great models are the second. And everything that these illustrious writers have left us of a philosophical or intellectual nature concerning the talent of the orator only proves how difficult it is to approach them. They were too enlightened to claim that they were showing the way; they doubtless wished merely to mark out its pitfalls. Those pedantic puerilities which have been honored by the name of Rhetoric, or rather which only serve to render that name ridiculous, are to the oratorical art what scholasticism is to true philosophy. They are capable of giving only the most false and barbarous idea of Eloquence. However, although people have begun to recognize rather universally the abuse of these puerilities, their long-established position as a distinguished branch of human knowledge does not yet permit us to banish them.[42] Perhaps, to the honor of our discernment, that time will one day come.

It is not enough for us to live with our contemporaries and to dominate them. Being animated by curiosity and self-esteem, we try, in our natural eagerness, to embrace the past, the present, and the future all at the same time. We wish simultaneously to live with those who will follow us and to have lived with those who have preceded us. From these [desires] come the origin and the study of History, which, while

---

[42] D'Alembert enlarged upon these thoughts in his stinging attack upon the sterile instruction in rhetoric in his controversial reforming article "Collège," *Encyclopédie*, III, 634–37. This article was highly offensive to the Jesuits, who rightly construed it as an attack on the schools of secondary education, which they practically monopolized in France.

uniting us with past centuries through the spectacle of their vices, their virtues, their knowledge, and their errors, transmits our own [virtues and defects] to the centuries of the future. It is from History that we learn to hold men in high regard solely for the good that they do and not for the imposing pomp which surrounds them. The sovereigns, those quite wretched men from whom everything conspires to hide the truth, can judge themselves ahead of time at this terrible and honest tribunal; the testimony of History toward those of their predecessors who resemble them is the image of posterity's judgment upon themselves.

Chronology and Geography are the two offshoots and supports of the science of which we are speaking. The one, Chronology, locates men in time, so to speak. The other, Geography, distributes them over our globe. Both draw considerable help from the history of the earth and of the heavens, that is, from historical facts and celestial observations. And if it were permitted here to borrow the language of the poets, one could say that the sciences of time and place are daughters of Astronomy and History.[43]

One of the principal rewards of the study of empires and their revolutions lies in the examination of how men, having been separated into various great families, so to speak, have formed diverse societies, how these different societies have given birth to different types of governments, and how they

[43] D'Alembert was the author of several articles concerning Chronology and its relationship to Astronomy in the *Encyclopedia*, e.g., "Calendrier," vol. II, and "Chronologie," vol. III. Later d'Alembert developed and enriched his conception of the nature and importance of the study of history. He became more and more explicit in his appreciation of the value of its philosophic study for the progress and understanding of the sciences. Also, in his later exposition, history becomes essential for the progress of the "empirical" study of human ethics. In his splendid article "Élémens des sciences," *Encyclopédie*, V, 495–96, he expounds the value of the study of history for the philosophic treatment of the moral universe. The philosopher, he says, regards history as "a collection of moral experiments made on mankind," which he collects, compares, and observes in order to discover the principles, just as a doctor compares and collects medical observations.

have tried to distinguish themselves both by the laws that they have given themselves and by the particular signs that each has created in order that its members might communicate more easily with one another. Such is the source of that diversity of languages and laws which has become an object of considerable study, to our misfortune. Such is also the origin of Politics, a sort of ethics of a particular and superior kind, to which the principles of ordinary ethics can on occasion be accommodated only with much subtlety. Penetrating into the essentials of the government of states, Politics distinguishes what can preserve, enfeeble, or destroy them. Perhaps this is the most difficult study of all, by virtue of the profound knowledge of peoples and men it demands and because of the extent and the variety of the talents that it presupposes. Such is the case especially when the man of politics endeavors not to forget that natural law, being anterior to all particular conventions, is thus the first law of peoples, and that to be a statesman one must not cease to be a man.

Such are the principal branches of that part of human knowledge which consists either in the direct ideas which we have received through our senses, or in the combination or comparison of these ideas—a combination which in general we call *Philosophy*.[44] These branches are subdivided into an infinite number of others, the enumeration of which would be enormously long, and belongs more to this work itself [the *Encyclopedia*] than to its preface.

Since the first operation of reflection consists in drawing together and uniting direct notions, we of necessity have begun this Discourse by looking at reflection from that point of view and reviewing the different sciences that result from it. But the notions formed by the combination of primitive ideas are not the only ones of which our minds are capable. There is

[44] For the encyclopedists Philosophy ceases to be a particular form of knowledge or reflection; it becomes truly the sum of reasoned human knowledge in all fields, and it is not far from being identified with the *Encyclopedia* itself. See Hubert's discussion of the subject in *Les Sciences sociales dans l'Encyclopédie*, p. 321.

another kind of reflective knowledge, and we must turn to it now. It consists of the ideas which we create for ourselves by imagining and putting together beings similar to those which are the object of our direct ideas. This is what we call the imitation of Nature, so well known and so highly recommended by the ancients. Since the direct ideas that strike us most vividly are those which we remember most easily, these are also the ones which we try most to reawaken in ourselves by the imitation of their objects. Although pleasant objects [of reality] have a greater impact on us because they are real rather than mere imitations, we are somewhat compensated for that loss of attractiveness [45] by the pleasure which results from imitation. As for the objects which, when real, excite only sad or tumultuous sentiments, imitation of them is more pleasing than the objects themselves, because it places us at precisely that distance where we experience the pleasure of the emotion without feeling its disturbance. That imitation of objects capable of exciting in us lively, vivid, or pleasing sentiments, whatever their nature may be, constitutes in general the imitation of *la belle Nature,* about which so many authors have written without presenting a clear idea of it.[46] They fail to do so either because *la belle Nature* can be perceived by only an extremely delicate sensitivity, or perhaps also because in this matter the limits which distinguish the arbitrary from the true are not yet well defined and leave some area open to opinion.

Painting and Sculpture ought to be placed at the head of that knowledge which consists of imitation, because it is in those arts above all that imitation best approximates the objects represented and speaks most directly to the senses. Architecture, that art which is born of necessity and perfected by luxury, can be added to those two. Having developed by de-

[45] Later editions: "what they lose of attractiveness in this latter case. . . ."

[46] D'Alembert's treatment of those areas of human thought that have to do with creative imagination, the fine arts, reflects a continuing preoccupation of eighteenth-century authors with the traditions of seventeenth-century classicism. (See n. 8, p. 67.)

grees from cottages to palaces, in the eyes of the philosopher it is simply the embellished mask, so to speak, of one of our greatest needs. The imitation of *la belle Nature* in Architecture is less striking and more restricted than in Painting or Sculpture. The latter express all the parts of *la belle Nature* indifferently and without restriction, portraying it as it is, uniform or varied; while Architecture, combining and uniting the different bodies it uses, is confined to imitating the symmetrical arrangement that Nature observes more or less obviously in each individual thing, and that contrasts so well with the beautiful variety of all taken together.[47]

Poetry, which comes after Painting and Sculpture, and which imitates merely by means of words disposed according to a harmony agreeable to the ear, speaks to the imagination rather than to the senses. In a touching and vivid manner it represents to the imagination the objects which make up this universe. By the warmth, the movement, and the life which it is capable of giving, it seems rather to create than to portray them. Finally, music, which speaks simultaneously to the imagination and to the senses, holds the last place in the order of imitation—not that its imitation is less perfect in the objects which it attempts to represent, but because until now it has apparently been restricted to a smaller number of images. This should be attributed less to its nature than to the lack of sufficient inventiveness and resourcefulness in most of those who cultivate it. It will not be useless to make some reflections on this subject. In its origin music perhaps was intended only to represent noise. Little by little it has become a kind of discourse, or even language, through which the different sentiments of the soul, or rather its different passions, are expressed. But why reduce this kind of expression to passions alone, and why not extend it as much as possible to the sensations them-

---

[47] *La belle Nature* is defined by Jaucourt *(Encyclopédie,* XI, 42) as nature embellished and perfected by the fine arts for use and pleasure. The artist makes a choice of the most beautiful parts of nature to form an exquisite whole, which would be more beautiful than nature itself. It is not direct imitation, but rather the representation of nature as it could be.

selves? Although the perceptions that we receive through various organs differ among themselves as much as their objects, we can nevertheless compare them according to another point of view which is common to them: that is, by the pleasurable or disquieting effect they have upon our soul. A frightening object, a terrible noise, each produces an emotion in us by which we can bring them somewhat together, and we can often designate both of these emotions either by the same name or by synonymous names. Thus, I do not see why a musician who had to portray a frightening object could not succeed in doing so by seeking in nature the kind of sound that can produce in us the emotion most resembling the one excited by this object. I say the same of agreeable sensations. To think otherwise would be to wish to restrict the limits of art and of our pleasures. I confess that the kind of depiction of which we are speaking here demands a subtle and profound study of the shadings which differentiate our sensations; thus it is not to be hoped that these shadings will be distinguished by an ordinary talent. Grasped by the man of genius, perceived by the man of taste, understood by the man of intelligence, they are lost on the multitude. Any music that does not portray something is only noise; and without that force of habit which denatures everything, it would hardly create more pleasure than a sequence of harmonious and sonorous words stripped of order and connection. It is true that a musician desirous of portraying everything would in many circumstances give us scenes of harmony which would not be grasped by vulgar senses. But all that can be concluded from this is that after having created an art of learning music one ought also to create an art of listening to it.[48]

[48] Rousseau complimented d'Alembert on his theory of musical imitation, which he found "très-juste et très neuve." *Correspondance générale de J.-J. Rousseau,* ed. Th. Dufour (Paris, 1924), letter of June, 1751, II, 11–12. However, Rulhière energetically attacked the same passage as being an example of what happened when geometers and scientists came to feel they were true judges of music because they understood the physics of sound (quoted in A. R. Oliver, *The Encyclopedists as Critics of Music* [New York, 1947], p. 94). D'Alembert's notions were perfectly typical of

We will stop enumerating the principal parts of our knowledge here. If one now looks at them all together and attempts to find some general points of view which can serve to differentiate them, one finds that some which are purely practical in nature [arts] have as their aim the execution of something. Others of a purely speculative nature [sciences] are limited to the examination of their object and the contemplation of its properties.[49] Finally, still others derive practical use from the speculative study of their object. Speculation and practice constitute the principal difference that distinguishes the *Sciences* from the *Arts,* and it is more or less according to this concept that we have given one or another name to each of the parts of our knowledge. We must acknowledge, however, that our ideas are not yet well established on this subject. Often we do not know what names to give most of those parts of knowledge in which speculation is united with practice, and it is still disputed in the schools, for example, whether Logic is an art or a science. The problem would soon be resolved by answering that it is simultaneously one and the other. How many questions and how much trouble we would spare ourselves if we finally determined the meaning of words in a clear and precise way!

In general the name *Art* may be given to any system of knowledge which can be reduced to positive and invariable rules independent of caprice or opinion. In this sense it would be permitted to say that several of our sciences are arts when they are viewed from their practical side. But just as there are rules for the operations of the mind or soul, there are also rules for those of the body: that is, for those operations which, applying exclusively to external bodies, can be executed by

---

those of the other encyclopedists who, as Oliver writes, "believed that all music was one and that its principal object was to 'paint.' . . . All this thinking grew out of the encyclopedists' lamentable attempt to force music into the imitation-of-nature thesis" (p. 61). See n. 56, p. 100, for a discussion of d'Alembert as music critic.

[49] D'Alembert makes the same distinction here that Diderot makes between an art and a science (see n. 5, p. 4).

hand alone. Such is the origin of the differentiation of the arts into liberal and mechanical arts, and of the superiority which we accord to the first over the second. That superiority is doubtless unjust in several respects. Nevertheless, none of our prejudices, however ridiculous, is without its reason, or to speak more precisely, its origin, and although philosophy is often powerless to correct abuses, it can at least discern their source. After physical force rendered useless the right of equality possessed by all men, the weakest, who are always the majority, joined together to check it. With the aid of laws and different sorts of governments they established an "inequality of convention" in which force ceased to be the defining principle.[50] Even though they united with good reason to preserve this "inequality of convention" once it was well established, men have not been able to resist complaining against it secretly because of that desire for superiority which nothing has been able to destroy in them. Thus, they have sought a sort of compensation in a less arbitrary inequality. Since physical force enchained by laws is no longer capable of offering any means of superiority, they have been reduced to seeking a principle of inequality in the difference of intellectual excellence—a principle which is equally natural, more peaceful, and more useful to society. Thus the most noble part of our being has in some measure taken vengeance for the first advantages which the basest part had usurped, and the talents of the mind have been generally recognized as superior to those of the body. The mechanical arts, which are dependent upon manual operation and are subjugated (if I may be permitted this term) to a sort of routine, have been left to those among men whom prejudices have placed in the lowest class. Poverty has forced these men to turn to such work more often than taste and genius have attracted them to it. Subsequently it became a reason for holding them in contempt—so much does poverty harm everything that accompanies it. With regard to the free operations

[50] This reference to the "social contract" adds one more fragment to the very sketchy treatment of political theory in the *Discourse* (see n. 20, p. 12).

of the mind, they have been apportioned to those who have believed themselves most favored of Nature in this respect. However, the advantage that the liberal arts have over the mechanical arts, because of their demands upon the intellect and because of the difficulty of excelling in them, is sufficiently counterbalanced by the quite superior usefulness which the latter for the most part have for us. It is this very usefulness which reduced them perforce to purely mechanical operations in order to make them accessible to a larger number of men. But while justly respecting great geniuses for their enlightenment, society ought not to degrade the hands by which it is served. The discovery of the compass is no less advantageous to the human race than the explication of its properties would be to physical science. Finally, considering in itself the principle of the distinction about which we are speaking, how many alleged scholars are there for whom science is in truth only a mechanical art? What real difference is there between a head stuffed with facts without order, without utility, and without connection, and the instinct of an artisan reduced to mechanical operation?

The contempt in which the mechanical arts are held seems to have affected to some degree even their inventors. The names of these benefactors of humankind are almost all unknown, whereas the history of its destroyers, that is to say, of the conquerors, is known to everyone. However, it is perhaps in the artisan that one must seek the most admirable evidences of the sagacity, the patience, and the resources of the mind. I admit that most of the arts have been invented only little by little and that it required a rather long sequence of centuries to bring watches, for example, to their present point of perfection. But is not the same true of the sciences? How many of the discoveries that have immortalized their authors had been prepared by the works of preceding centuries, sometimes being already brought to their maturity, to the point where they required just one step more to be accomplished? And not to leave watchmaking, why are not those to whom we owe the fusee, the escapement, the repeating-works [51] of watches equally esteemed with those who have worked successively to perfect algebra? Moreover, if I may believe a few philosophers who have not been deterred from studying the arts by the prevailing contempt for them, there are certain machines that are so complicated, and whose parts are all so dependent on one another, that their invention must almost of necessity be due to a single man. Is not that man of genius, whose name is shrouded in oblivion, well worthy of being placed beside the small number of creative minds who have opened new routes for us in the sciences?

Among the liberal arts that have been reduced to principles, those that undertake the imitation of Nature have been called the Fine Arts because they have pleasure for their principal object.[52] But that is not the only characteristic distinguishing them from the more necessary or more useful liberal arts, such as Grammar, Logic, and Ethics. The latter have fixed and settled rules which any man can transmit to another, whereas the practice of the Fine Arts consists principally in an invention which takes its laws almost exclusively from genius. The rules which have been written concerning these arts are, properly speaking, only the mechanical part. Their effect is somewhat like that of the telescope; they only aid those who see.

From everything we have said heretofore it follows that the different ways in which our mind operates on objects and also the different uses which it derives from these objects are the first available means of generally distinguishing the various parts of knowledge from one another. Everything is related to our needs, whether from absolute necessity, or from convenience and pleasure, or even from custom and caprice. The more the needs are remote or difficult to satisfy, the slower the knowledge intended to satisfy them will be in making its ap-

[51] These are all technical watchmakers' terms whose definitions can be found in the *Encyclopedia* or in an unabridged dictionary.

[52] Eloquence, poetry, music, painting, sculpture, architecture, and engraving are in this category.

pearance. Think of the progress Medicine would have made at the expense of the purely speculative sciences if it were as certain as Geometry. But there are still other very marked characteristics in the way our knowledge affects us and in the different judgments that our soul makes of ideas. These judgments are designated by the words evidence, certitude, probability, feeling, and taste.

Evidence properly pertains to the ideas whose connection the mind perceives immediately.[53] Certitude pertains to those whose connection can be known only by the aid of a certain number of intermediate ideas, or, what is the same thing, to propositions whose identity with a self-evident principle can be discovered only by a circuit of greater or lesser length. From this it would follow that according to the nature of minds, what is evident for one person would sometimes only be certain for another. Taking the words in another sense, it could also be said that evidence is the result of the operations of the mind alone and is related to metaphysical and mathematical speculations;[54] and that certitude is more appropriate to physical objects, the knowledge of which is the fruit of the constant and invariable testimony of our senses. Probability applies principally to historical facts and generally to all past, present, and future events that we attribute to a kind of chance because we cannot perceive the causes of them. The part of that knowledge which has the present and the past for its object often produces a conviction as strong as that which comes from axioms, although it is founded merely on testimony. Feeling is of two sorts. The one concerned with moral truths is called conscience. It is a result of natural law and of our conception of good and evil.[55] One could call it evi-

[53] The word "evidence" as used here may be defined as the condition of being self-evident.

[54] Later editions: "operations."

[55] "Natural law," as we have seen, means for d'Alembert the natural fact that men in society, in their urge for self-preservation and freedom from pain, have always naturally resisted oppression. This fact of human experience is, of course, very close to the level of feeling. The fact of

dence of the heart, for, although it differs greatly from the evidence of the mind which concerns speculative truths, it subjugates us with the same force.[56] The other sort of feeling pertains in particular to the imitation of *la belle Nature* and to what we call beauties of expression. It grasps sublime and striking beauties with rapture, subtly discerns hidden beauties, and proscribes those that merely feign their appearance. Often, indeed, it pronounces severe judgments without bothering to describe in detail the motives for them, because these motives depend upon a multitude of ideas that are difficult to expound all at once and still more difficult to transmit to others. It is to this kind of feeling that we owe taste and genius, which are distinguished from one another in that genius is the feeling that creates and taste the feeling that judges.

After reviewing the different parts of our knowledge and the characteristics that distinguish them, it remains for us only to make a genealogical or encyclopedic tree which will gather the various branches of knowledge together under a single point

---

oppression and the urge for self-preservation and freedom from pain seems in d'Alembert's mind to have stimulated an almost automatic reaction in the feelings of human beings that responds directly to right and wrong actions—conscience. See n. 20, p. 12, and our Introduction, p. xxviii.

[56] The anonymous author of the long review of the *Discourse* in the *Journal des Sçavans*, October, 1751, p. 220, took exception to this statement in the following terms:

> And [in the *Discourse*] ethical truths have only an evidence of the heart, founded on the sentiment of the conscience, and completely different from the evidence of the intelligence attributed to speculative truths. Are not the first principles of ethics of an evidence recognized by the light of Reason? Do men have no rule for their action other than conscience? Has M. d'Alembert realized the frightful consequences of such a principle?

D'Alembert defended himself in a later edition of the *Discourse* for having, as he wrote, "admitted with Pascal the existence of some truths which come to the heart and others which, without being in conflict, come to the mind." (*Œuvres*, I, 15). Besides, he seems to say here that conscience is shaped at least in part by our intellectual conception of good and evil.

of view and will serve to indicate their origin and their relationships to one another. We will explain in a moment the use to which that tree may be put according to our claims, but the execution itself is not without difficulty. Although the philosophical history we have just given of the origins of our ideas is very useful in facilitating such a work, it should not be thought that the encyclopedic tree ought to be, or even can be, slavishly subject to that history. The general system of the sciences and the arts is a sort of labyrinth, a tortuous road which the intellect enters without quite knowing what direction to take. Impelled, first of all, by its needs and by those of the body to which it is united, the intelligence studies the first objects that present themselves to it. It delves as far as it can into the knowledge of these objects, soon meets difficulties that obstruct it, and whether through hope or even through despair of surmounting them, plunges on to a new route; now it retraces its footsteps, sometimes crosses the first barriers only to meet new ones; and passing rapidly from one object to another, it carries through a sequence of operations on each of them at different intervals, as if by jumps. The discontinuity of these operations is a necessary effect of the very generation of ideas. However philosophic this disorder may be on the part of the soul,[57] an encyclopedic tree which attempted to portray it would be disfigured, indeed utterly destroyed.

Moreover, as we have already shown in the subject of Logic, most of the sciences which we regard as including the basic principles of all the others, and which ought for this reason to occupy the first places in the encyclopedic arrangement, do not have the first places in the genealogical arrangements of ideas because they were not invented first. Indeed, in the beginning we [human beings] of necessity studied individual things. It is only after having considered their particular and palpable properties that we envisaged their general and common properties and created Metaphysics and Geometry by intellectual abstraction. Only after the long usage of the first

[57] Later editions: *"l'esprit"* instead of *"l'âme."*

signs have we perfected the art of these signs to the point of making a science of them. And it is only after a long sequence of operations on the objects of our ideas that, through reflection, we have at length given rules to these operations themselves.

Finally, the system of our knowledge is composed of different branches, several of which have a common point of union. Since it is not possible, starting out from this point, to begin following all the routes simultaneously, it is the nature of the different minds that determines which route is chosen. Rarely does a single mind travel along a large number of these routes at the same time. In the study of Nature, men at first applied themselves, as if in concert, to satisfying the most pressing needs. But when they came to less absolutely necessary knowledge, they were obliged to divide it among themselves, and each one moved forward in almost equal step with the others. Thus several sciences have been contemporaneous, so to speak. But when tracing in historical order the progress of the mind, one can only embrace them successively.

It is not the same with the encyclopedic arrangement of our knowledge. This consists of collecting knowledge into the smallest area possible and of placing the philosopher at a vantage point, so to speak, high above this vast labyrinth, whence he can perceive the principal sciences and the arts simultaneously. From there he can see at a glance the objects of their speculations and the operations which can be made on these objects; he can discern the general branches of human knowledge, the points that separate or unite them; and sometimes he can even glimpse the secrets that relate them to one another. It is a kind of world map which is to show the principal countries, their position and their mutual dependence, the road that leads directly from one to the other. This road is often cut by a thousand obstacles, which are known in each country only to the inhabitants or to travelers, and which cannot be represented except in individual, highly detailed maps. These individual maps will be the different articles of the *Encyclo-*

pedia and the Tree or Systematic Chart will be its world map.[58]

But as, in the case of the general maps of the globe we inhabit, objects will be near or far and will have different appearances according to the vantage point at which the eye is placed by the geographer constructing the map, likewise the form of the encyclopedic tree will depend on the vantage point one assumes in viewing the universe of letters. Thus one can create as many different systems of human knowledge as there are world maps having different projections, and each one of these systems might even have some particular advantage possessed by none of the others. There are hardly any scholars who do not readily assume that their own science is at the center of all the rest, somewhat in the way that the first men placed themselves at the center of the world, persuaded that the universe was made for them. Viewed with a philosophical eye, the claim of several of these scholars could perhaps be justified by rather good reasons, quite aside from self-esteem.

In any case, of all the encyclopedic trees the one that offered the largest number of connections and relationships among the sciences would doubtless deserve preference. But can one flatter oneself into thinking it has been found? We cannot repeat too often that nature is composed merely of individual things which are the primary object of our sensations and di-

---

[58] The explanation of the chart begins on p. 143. This section was anticipated in the following passage cited by Lovejoy (p. 232) from Thomas Sprat's *History of the Royal Society* (1667), p. 110:

> Such is the dependence amongst all the orders of creatures; the animate, the sensitive, the rational, the natural, the artificial; that the apprehension of one of them, is a good step toward the understanding of the rest. And this is the highest pitch of humane reason: to follow all the links of this chain, till all their secrets are open to our minds; and their works advanc'd or imitated by our hands. This is truly to command the world: to rank all the varieties and degrees of things so orderly upon one another, that standing on the top of them, we may perfectly behold all that are below; and make them all serviceable to the quiet and peace and plenty of Man's life. . . .

rect perceptions. To be sure, we note in these individual things common properties by which we compare them and dissimilar properties by which we differentiate them. And these properties, designated by abstract names, have led us to form different classes in which these objects have been placed. But often such an object, which because of one or several of its properties has been placed in one class, belongs to another class by virtue of other properties and might have been placed accordingly. Thus, the general division remains of necessity somewhat arbitrary. The most natural arrangement would be the one in which the objects followed one another by imperceptible shadings which serve simultaneously to separate them and to unite them.[59] But the small number of beings known to us does not permit us to indicate these shadings. The universe is but a vast ocean, on the surface of which we perceive a few islands of various sizes, whose connection with the continent is hidden from us.

One could construct the tree of our knowledge by dividing it into natural and revealed knowledge, or useful and pleasing knowledge, or speculative and practical knowledge, or evident, certain, probable, and sensitive knowledge, or knowledge of things and knowledge of signs, and so on into infinity. We have chosen a division which has appeared to us most nearly satisfactory for the encyclopedic arrangement of our knowledge and, at the same time, for its genealogical arrangement.

---

[59] This paragraph seems to be a direct application of Buffon's definition of biological continuity to the entire range of encyclopedic knowledge. It appears to be based, in much the same vocabulary, on Buffon's "Premier Discours" to his *Histoire naturelle*, whose first three volumes appeared in 1749 (*Œuvres philosophiques de Buffon* [Paris, 1954], I, 7-26), and it is one of the most striking expressions in the eighteenth century of the notion of continuity described by Lovejoy in *The Great Chain of Being*. D'Alembert returns to the subject in the article "Cosmologie" in the *Encyclopedia* and in his *Élémens de philosophie* (1759). The function of the philosopher, according to him, is to add new links and to fill in the gaps between them as much as possible so that our knowledge can be made ever more useful in improving the human condition.

We owe this division to a celebrated author [Bacon] of whom we will speak later in this preface.[60] To be sure, we have thought it necessary to make some changes in his division, of which we will render an account; but we are too aware of the arbitrariness which will always prevail in such a division to believe that our system is the only one or the best. It will be sufficient for us if our work is not entirely disapproved of by men of intelligence. We do not wish to resemble that multitude of naturalists (censured with such good reason by a modern philosopher) whose energies have been ceaselessly devoted to dividing the productions of Nature into genera and species, consuming an amount of time in this labor which would have been employed to much better purpose in the study of those productions themselves.[61] What would be said of an architect, who, having to build an immense edifice, passed his whole life in drawing the plans for it? Or likewise what would we say of an inquisitive person who, proposing to inspect an enormous palace, spent all his time in observing the entryway?

The objects to which our soul applies itself are either spiritual or material, and our souls are occupied with these objects either through direct ideas or through reflective ideas. The system of direct knowledge consists simply in the purely passive and almost mechanical collection of this same knowledge; this is what we call memory. Reflection is of two kinds (as we have already observed): either it reasons on the objects of direct ideas, or it imitates them. Thus memory, reason (strictly speaking), and imagination are the three different manners in which our soul operates on the objects of its thoughts. We do not take imagination here to be the ability to represent objects to oneself, since that faculty is simply the memory itself of sensible objects, a memory which would be continually in action if it were not assisted and relieved by the

---

[60] Later editions: "this Discourse."

[61] In his *Premier Discours* Buffon makes a sweeping attack on the great systematizers of biology in the seventeenth and eighteenth centuries, particularly Linnaeus and Tournefort. The system of nomenclature of the latter was generally adopted in the *Encyclopedia*.

invention of signs. We take imagination in the more noble and precise sense, as the talent of creating by imitating.

These three faculties form at the outset the three general divisions of our system of human knowledge: History, which is related to memory; Philosophy, which is the fruit of reason; and the Fine Arts, which are born of imagination. Placing reason ahead of imagination appears to us to be a well-founded arrangement and one which is in conformity with the natural progress of the operations of the mind. Imagination is a creative faculty, and the mind, before it considers creating, begins by reasoning upon what it sees and knows. Another motive which should decide us to place reason ahead of imagination is that in the latter faculty the other two are to some extent brought together. The mind creates and imagines objects only insofar as they are similar to those which it has known by direct ideas and by sensations. The more it departs from these objects, the more bizarre and unpleasant are the beings which it forms. Thus, in the imitation of Nature, invention itself is subjected to certain rules. It is principally these rules which form the philosophical part of the Fine Arts, which is still rather imperfect because it can be the work only of genius, and genius prefers creation to discussion.

Finally, if we examine the progress of reason in its successive operations, we will again agree that it ought to precede imagination in the arrangement of our faculties, because reason in a way leads to imagination by the last operations which it makes on objects. These operations consist entirely in "creating" general beings, so to speak, which no longer fall within the immediate competence of our senses since they are separated from their subject by abstraction. Thus of all the sciences that pertain to reason, Metaphysics and Geometry are those in which imagination plays the greatest part. I ask pardon of those superior wits who are detractors of Geometry; doubtless they do not think [62] themselves so close to it, although all that separates them perhaps is Metaphysics. Imagination acts no less in a geometer who creates than in a poet

---

[62] Later editions: ". . . they did not think. . . ."

who invents. It is true that they operate differently on their object. The first shears it down and analyzes it, the second puts it together and embellishes it. It is true, further, that these different ways of operating stem from different sorts of minds, and for this reason the talents of a great geometer and those of a great poet will perhaps never be found together.[63] But whether or not they are mutually exclusive, they have no right to hold one another in contempt. Of all the great men of antiquity, Archimedes is perhaps the one who most deserves to be placed beside Homer. I hope that this digression by a geometer who loves his art will be pardoned, and that he will not be accused of being an excessive enthusiast; and I return to my subject.

The general distribution of beings into spiritual and material provides a subdivision of the three general branches. History and Philosophy are occupied with each of these two kinds of beings, while imagination deals only with purely material beings, which is a new reason for placing it last in the arrangement of our faculties. At the head of the spiritual beings is God, who necessarily holds the first rank by virtue of His nature and of our need to know Him. Below that Supreme Being are the created spiritual beings whose existence is taught us by Revelation. Next comes man. Composed of two principles, he belongs by virtue of his soul to the spiritual beings and by virtue of his body to the material world. And finally comes that vast universe which we call the corporeal world, or Nature. We do not know why the celebrated author [Bacon] who serves as our guide in this arrangement has placed Nature before man in his system. It seems, on the contrary, that everything engages us to put man in the passageway that separates God and the spiritual beings from material bodies.

Insofar as it is related to God, History includes either Rev-

[63] This passage is largely to justify altering Bacon's order of arrangement of knowledge which placed imagination before reason. Note that "geometer" in d'Alembert's definition is a term that includes all mathematicians and is not strictly limited to practitioners of geometry alone. (See the article "Géometre," *Encyclopédie*, VII, 627.)

elation or tradition, and according to these two points of view, is divided into sacred history and ecclesiastical history. The history of man has for its object either his actions or his knowledge, and consequently is civil or literary. In other words, it is divided between the great nations and the great geniuses, between the kings and the men of letters, between the conquerors and the philosophers. Finally, the history of Nature is the history of the innumerable productions that we observe therein, forming a quantity of branches almost equal in number to those diverse productions. Among these different branches, a distinguished place should be given to the history of the arts, which is simply the history of the use which men have made of the productions of Nature to satisfy their needs or their curiosity.

Such are the principal objects of memory. Let us turn now to the faculty that reflects and reasons. Both the spiritual and the material beings on which that faculty acts have some general properties such as existence, possibility, and duration. The examination of these properties constitutes at the outset that branch of Philosophy from which all others in part borrow their principles and which is called Ontology, or the science of being, or general Metaphysics. We descend from there to the different particular beings, and the science of these different beings is divided according to the same plan as that of History.

The science of God, called Theology, has two branches: Natural Theology has only such knowledge of God as reason unaided produces, a knowledge which is not of very great extent. Revealed Theology draws a much more perfect knowledge of that Supreme Being from sacred history. From this same Revealed Theology results the science of created spiritual beings. Here again we have felt we ought to depart from our author [Bacon]. It seems to us that science, considered as belonging to reason, ought not to be divided into Theology and Philosophy as it has been by him. For Revealed Theology is simply reason applied to revealed facts. One can say that it belongs to History by virtue of the dogma that it teaches and to Philosophy by virtue of the consequences that it draws from these dogmas. Thus, to separate Theology from Philosophy would be to cut the offshoot from the trunk to which it is united by its very nature. It seems also that the science of spiritual beings is bound much more closely to Revealed Theology than to Natural Theology.[64]

The first part of the science of man is that of the soul, and that science has for its aim either the speculative knowledge of the human soul or knowledge of its operations. Speculative knowledge of the soul derives in part from Natural Theology and in part from Revealed Theology, and is called Pneumatology or Particular Metaphysics. The knowledge of its operations is subdivided into two branches, these operations being capable of having either the discovery of truth or the practice of virtue for their object. The discovery of truth, which is the aim of Logic, produces the art of transmitting it to others. Thus, the use that we make of Logic is partly for our own advantage, partly for that of others of our species. The rules of Ethics are less related to isolated man and necessarily presume that he is in society with other men.

The science of Nature is simply the science of bodies. But since bodies have general properties which are common to them, such as impenetrability, mobility, and extension, the science of Nature ought therefore to begin with the study of these properties. They have, so to speak, a purely intellectual side, by which they open an immense scope to the speculations of the mind, and a material and sensible side by which we can measure them. Intellectual speculation is related to General Physics, which is, properly speaking, simply the metaphysics of bodies, and measurement is the object of Mathematics, whose divisions extend almost to infinity.

These two sciences lead to Particular Physics, which studies the bodies in themselves and whose sole object is individual things. Our own body ought to hold first rank among the

[64] D'Alembert was taken to task in the *Journal des Sçavans*, October, 1751, pp. 222–23, for subordinating Theology to Philosophy and for having degraded the study of natural religion. He defended himself in the later editions of the *Discourse*.

bodies whose properties it is worthwhile for us to know, and it is immediately followed by those which we most need to know for self-preservation. Whence result Anatomy, Agriculture, Medicine, and their different branches. Finally, all the natural bodies submitted to our examination produce the innumerable other parts of reasoned Physics.

Painting, Sculpture, Architecture, Poetry, Music, and their different divisions make up the third general distribution, which is born of imagination and whose parts are comprised under the name of Fine Arts. We can also include them under the general title of Painting [portrayal], because all the Fine Arts can be reduced to that and differ only by the means which they use. Finally, we could relate them all to Poetry by taking this word in its natural signification, which is simply invention or creation.

Such are the principal parts of our encyclopedic tree. They will be found in more detail at the end of this Preliminary Discourse. We have made a sort of chart of them to which we have joined a much more extended explication than has just been given. This chart and this explication have already been published in the *Prospectus* in order to sound out the pleasure of the public. We have made some changes which will be easy to recognize. They are the fruit either of our reflections or of the counsels of a few philosophers who have been sufficiently public-spirited to take an interest in our work. If the enlightened public gives its approbation to these changes, it will be the reward for our tractableness, and if it does not approve them, we will only be more strongly convinced of the impossibility of designing an encyclopedic tree that would please everyone.

There is this advantage in the general division of our knowledge according to our three faculties: it could also provide the three divisions of the literary world into Scholars, Philosophers, and *beaux esprits*,[65] so that, after having designed the tree of sciences, one would be able to construct the tree of men

---

[65] "*Beaux esprits*" in this context could be rendered "Creative Artists." It includes writers, composers, painters, and practitioners of the various arts of imagination.

of letters on the same pattern. Memory is the talent of the first [group]; wisdom belongs to the second; and the last have pleasure as their portion. Thus, if one considers memory as a beginning of reflection and adds to it the reflection that combines and the one that imitates, one could say in general that the differences which exist among men are determined by the variation in the number and nature of each man's reflective ideas. One could also say that reflection, taken in the most extended sense that one can give it, molds the character of the intelligence and distinguishes its different types. For the rest, the three kinds of "republics" into which we have just distributed the men of letters ordinarily have nothing in common, except the lack of esteem in which they hold one another. The poet and the philosopher treat each other as madmen who feed on fancies. Both regard the scholar as a sort of miser who thinks only of amassing without enjoying and who indiscriminately heaps up the basest metals along with the most precious. And the scholar, who considers everything which is not fact to be idle words, holds the poet and the philosopher in contempt as being men who think they are rich because their expenses exceed their resources.

It is thus that people avenge themselves for advantages they do not have. Men of letters would better understand their interests, if, instead of trying to isolate themselves, they recognized the reciprocal need they have of each other's works and the assistance which they could draw from them. Undoubtedly society owes its principal enjoyments to *beaux esprits* and its enlightenment to philosophers, but neither of them appreciates how much they owe to memory. Memory includes the primary material of all our knowledge, and the works of the scholars have often furnished the subjects upon which the philosopher and the poet exercise their skills. When the ancients called the Muses the daughters of memory (a modern author has said), they appreciated perhaps how much that faculty of our mind is necessary to all the others. The Romans raised temples to her, just as they did to Fortune.

It remains for us to show how we have tried to reconcile the encyclopedic arrangement with the alphabetical arrangement in this Dictionary. To accomplish this task we have employed three means: the chart at the beginning of the work, the [designation of the] science to which each article is related, and the manner in which the article is treated. Ordinarily the name of the science to which the article belongs has been placed after the word that constitutes the subject of the article. Simply by referring to the chart one can see what rank this science occupied and hence understand the place that the article is to have in the Encyclopedia. If it happens that the name of the science is omitted, a reading of the article will suffice to make clear the science to which it is related, and even if we forget to point out, for example, that the word *Bomb* belongs to the military art, and the name of a city or country to geography, we have enough confidence in the intelligence of our readers to hope that they will not be shocked by such an omission. Moreover, through the arrangement of the contents of each article, especially in those of some length, it will hardly be possible to avoid seeing that such and such an article is related to another article, which belongs to a different science, and which in turn is related to a third article, and so forth. By means of the precision and frequency of the references to other articles [*les renvois*], we have tried to leave nothing to be desired on that score. For such references in this Dictionary are unusual in that they serve principally to indicate the connection of the materials, whereas in other works of this type, they are intended only to elucidate one article by another. Often, indeed, we have omitted the reference to another article because the terms of art or science which it would have designated are explained in individual articles which the reader will find by himself. Especially in the general articles on the sciences, we have tried to explain the aid which they give one another. Thus, three things make up the encyclopedic arrangement: the name of the science to which the article belongs, the position of that science in the tree, and the connection of the article with others in the same science or in a different science. This connection is indicated by the references to other articles or is easy to understand by means of the technical terms explained in their alphabetical place. We are not concerned here with the reasons which have made us prefer the alphabetical arrangement in this work to all others; these will be explained later when we speak of this collection as a Dictionary of Sciences and Arts.

For the rest, we will make two observations concerning the part of our work which involves the encyclopedic arrangement and which is intended more for enlightened men than for the multitude. The first is that it would often be absurd to expect to find an immediate connection between one article of this Dictionary and another article taken at random. Thus one would look in vain to find by what secret bonds the article *Conic Section* can be related to the article *Accusative*. The encyclopedic arrangement does not suppose that all the sciences stem from one another. They are branches which grow out of the same trunk, that is, out of the human intellect. These branches often have no immediate connection among themselves, and several are united only by the trunk itself. Thus, *Conic Section* concerns geometry, geometry leads to particular physics, this latter to general physics, and general physics to metaphysics. Metaphysics is very close to grammar, to which the word *accusative* belongs. But when one has arrived at this last point by the route we have just indicated, one is so far from the starting point that it has been completely lost from sight.

Our second remark is that one should not attribute more advantages to our encyclopedic tree than we claim to give it. The general divisions are useful in that they gather together a rather large number of objects, but it is not to be believed that this collection can replace the study of these objects themselves. It is a kind of enumeration of the knowledge that can be acquired—a frivolous enumeration for whoever would wish to let it go at that alone, but useful for whoever desires to go further. A single reasoned article on a particular object of science or art includes more material substance than all the possible divisions and subdivisions of the general terms. And, not to leave the comparison that we have made above with geographical maps, anyone who relied solely upon the encyclopedic tree for all his knowledge would hardly know more than a man who flattered himself that he knew the different peoples who inhabit the globe and the particular states which constitute it because he had acquired a general idea of the globe and its principal parts through world maps. Above all, it ought not to be forgotten in considering our systematic chart that the encyclopedic order which it presents is very different from the genealogical order of the operations of the mind. Nor should one forget that the sciences which are concerned with general beings are useful only insofar as they lead to those which have particular beings as their object, that only these particular beings truly exist, and that if our minds have created general beings, their purpose is to enable one to study successively the properties which, by their nature, exist at the same time in the same substance and cannot physically be separated. These reflections are the due fruit and result of what we have said up until now. And thus it is with them that we will end the first part of this Discourse.

[PART II]

We are now going to consider this work as a *Reasoned Dictionary of the Sciences and the Arts*. This object is all the more important, since it may possibly be of greater interest to the majority of our readers [than the object of Part I], and fulfilling it has required the most care and work. But before entering upon this subject in all the detail which may legitimately be required of us, it will be worth our while to examine at some length the present state of the sciences and the arts, and to show what series of steps have led us to it. Our metaphysical analysis of the origin and connection of the sciences has been of great utility in designing the encyclopedic tree; an historical analysis of the order in which our knowledge has developed in successive steps will be no less useful in enlightening us concerning the way we ought to convey this knowledge to our readers. Moreover, the history of the sciences is naturally bound up with that of the small number of great geniuses whose works have helped to spread enlightenment among men. And since these works have been helpful in a general way to our own, we ought to begin by speaking of them before giving account of the particular contributions and assistance that we have obtained. So as not to go back too far, let us turn our attention to the renaissance of letters.

When we consider the progress of the mind since that memorable epoch, we find that this progress was made in the sequence it should naturally have followed. It was begun with erudition, continued with belles-lettres, and completed with philosophy. This sequence differs, it is true, from that which a man would necessarily follow if left to his own intelligence or limited to exchanges with his contemporaries—a sequence that we analyzed in the first part of this Discourse.[1] For indeed we have demonstrated that a mind left in isolation would

[1] That sequence is Memory, Reason, and Imagination. See pp. 50–51.

of necessity encounter philosophy before it arrived at belles-lettres; whereas the regeneration of ideas, if we can speak thus, must necessarily have been different from their original generation, as it involved moving out of a long interval of ignorance that had been preceded by centuries of enlightenment. We are going to try to demonstrate this.

The masterpieces that the ancients left us in almost all genres were forgotten for twelve centuries. The principles of the sciences and the arts were lost, because the beautiful and the true, which seem to show themselves everywhere to men, are hardly noticed unless men are already apprised of them. Not that these unfortunate times were less fertile than others in rare geniuses; Nature is always the same.[2] But what could these great men do, scattered as they always are, from place to place, occupied with different purposes, and left to their solitary enlightenment with no cultivation of their abilities? Ideas which are acquired from reading and from association with others are the germ of almost all discoveries. It is like the air one breathes without thinking about it, to which one owes life; and the men of whom we are speaking were deprived of such sustenance. They were like the first creators of the sciences and the arts who have been forgotten because of their illustrious successors, and who, had they but come later, would themselves have caused the memory of the others to fade. The man who first discovered wheels and pinions would have invented watches in another century. Gerbert, situated in the time of Archimedes, would perhaps have equaled him.[3]

Most of the superior intelligences of those dark times called themselves poets or philosophers. And indeed what did it cost

[2] D'Alembert reproduces the famous assertion of Fontenelle concerning the continuity of the forces of nature, upon which he based his arguments in favor of the Moderns in his *Digression of the Ancients and the Moderns* (1688), in *Œuvres de Fontenelle* (Amsterdam, 1764), IV, 114.

[3] Gerbert of Aurillac (940?–1003) became Pope Sylvester II. He was a great medieval French scholar whose special competence was in mathematics and the physical sciences. This accounts for d'Alembert's admiration for him alone among the intellectuals of the Middle Ages, and it is for this reason he is coupled with Archimedes.

them to usurp two titles with which people bedeck themselves at so little expense, ever flattering themselves that they can hardly owe them to borrowed wit? They thought it useless to seek models for poetry in the works of the Greeks and the Romans, whose language was no longer spoken; and they mistook for the true philosophy of the ancients a barbarous tradition which disfigured it.[4] Poetry for them was reduced to a puerile mechanism. The careful examination of Nature and the grand study of mankind were replaced by a thousand frivolous questions concerning abstract and metaphysical beings—questions whose solution, good or bad, often required much subtlety, and consequently a great abuse of intelligence. Added to this confusion were the condition of slavery into which almost all of Europe was plunged and the ravages of superstition which is born of ignorance and which spawns it in turn. If one considers all these difficulties it will be plain that nothing was lacking to the obstacles that for a long time delayed the return of reason and taste. For liberty of action and thought alone is capable of producing great things, and liberty requires only enlightenment to preserve itself from excess.

And so one of those revolutions which make the world take on a new appearance was necessary to enable the human species to emerge from barbarism. The Greek [Byzantine] empire was destroyed, and its ruin caused the small remainder of knowledge in the world to flow back into Europe. The invention of printing and the patronage of the Medici and of Francis I revitalized minds, and enlightenment was reborn everywhere.[5]

[4] D'Alembert refers here to the much-hated scholasticism.

[5] An historian cannot but raise an objection to d'Alembert's gross and inaccurate explanation for the origin of the Renaissance, as well as his violent treatment of the Middle Ages. The fall of Constantinople came well after the first masterpieces of the Renaissance had appeared. Voltaire gave precisely the same description of the beginning of the Renaissance in his *Siècle de Louis XIV* (1751), in *Œuvres*, XIV, 155–56, and in his *Essai sur les mœurs*, in *Œuvres*, XI, 162. D'Alembert sensibly qualified the "fall of Constantinople" theory of the Renaissance in his article "Erudition," *Encyclopédie*, V, 915.

People turned first to the study of languages and history, which had perforce been abandoned during the centuries of ignorance. On emerging from barbarism, the human mind found itself in a sort of infancy. It was eager to accumulate ideas, but incapable at first of acquiring those of a higher order because of the kind of sluggishness in which the faculties of the soul had for so long a time been sunk. Of all these faculties, memory was the one which was cultivated first, because it is the easiest to satisfy and because the knowledge that is obtained with its help can be built up most easily. Thus they did not begin by studying Nature as the first men had had to do. They enjoyed an advantage which the earliest men lacked: they had the works of the ancients, which printing and the generosity of men of power and noble birth began to make common. They thought they needed only to read in order to become learned; and it is far easier to read than to understand. And so they devoured indiscriminately everything that the ancients left us in each genre. They translated them, commented on them, and out of a kind of gratitude began to worship them, although they were far from knowing their true worth.

These circumstances gave rise to that multitude of erudite men, immersed in the learned languages to the point of disdaining their own, who knew everything in the ancients except their grace and finesse, as a celebrated author has said, and whose vain show of erudition made them so arrogant because the cheapest advantages are rather often [6] those whose vulgar display gives most satisfaction. They acted like great lords who do not resemble their forefathers in any real merit but who are excessively proud of their ancestry. Moreover, that vanity was not without some degree of plausibility. The realm of erudition and of facts is inexhaustible; the effortless acquisitions made in it lead one to think that one's substance is continually growing, so to speak. On the contrary, the realm of reason and of discoveries is rather small. Through study in that realm, men often succeed only in unlearning what they

[6] Later editions: "ordinarily."

thought they knew, instead of learning what they did not know. That is why a scholar of most unequal merit must be much more vain than a philosopher or even perhaps a poet. For the inventive mind is always dissatisfied with its progress because it sees beyond, and for the greatest geniuses, even their self-esteem may harbor a secret but severe judge whom flattery may momentarily silence but never corrupt. Thus we should not be surprised that the scholars of whom we speak gloried so proudly in practicing a science that was thorny, often ridiculous, and sometimes barbarous.[7]

It is true that our century, which believes itself called upon to change every type of law and to render equitable judgments, does not think very well of these men who were formerly so celebrated. Belittling them is nowadays considered the proper thing to do; indeed many men pride themselves on it. We seem to be trying with our contempt to punish these scholars for their excessive self-esteem or for the unenlightened approbation of their contemporaries; and in trampling these idols under foot it appears we would wish to cause their very names to be forgotten. But any excess is unjust. Let us instead gratefully make use of the work of these industrious men. While making available to us everything that could be useful from the works of the ancients, it was inevitable that they would turn up what was useless also. We would not be able to mine gold without extracting many baser or less precious materials at the same time; these men would have separated out the good from the bad just as we do, if they had come later. Erudition was therefore necessary to lead us to belles-lettres.

Indeed, it was not necessary to read the writings of the ancients for long to be convinced that there were better things to learn in those very works in which formerly only facts or words were sought. Men soon perceived the beauties that the ancients had lavished upon them; for, if men need to be apprised of what is true, as we have stated above, in compensa-

[7] D'Alembert refers to the Latin and Greek scholarship of the humanists, like Budé of the French Renaissance.

tion, that is all they need. Their former admiration for the ancients could not have been livelier; now it became more judicious. However, it was still far from being wise. They believed that the only way to imitate the ancients was to copy them slavishly and that it was possible to speak well only in an ancient tongue. They did not realize that the study of words is a kind of passing inconvenience, necessary insofar as it facilitates the study of things, but becoming a real evil whenever it retards that study; that therefore their endeavor should have been restricted to familiarizing themselves with the Greek and Roman authors so as to profit from their finest thoughts; and that the trouble it took to write in an ancient language was just so much time lost from the advancement of reason. Furthermore, they did not see that if there were great beauties of style in the ancients that are lost to us, for the same reason there must also be many defects in them which escape our notice and which we run the risk of copying, mistaking them for beauties. And lastly, they did not realize that the best for which one could hope from the slavish use of ancient languages would be a weirdly diverse style conglomerated from an infinite number of different ones. This composite style would be correct and even admirable in the view of our moderns, but Cicero or Virgil would have found it ridiculous. In the same way we would laugh at a work in our language in which the author had collected together sentences from Bossuet, La Fontaine, La Bruyère, and Racine, quite correctly persuaded that, taken individually, each of these writers is an excellent model.

Such prejudice on the part of the first scholars produced in the sixteenth century a multitude of Latin poets, orators, and historians whose works, we must confess, too often draw their principal merit from a latinity which we can hardly evaluate. Some of these works can be compared to the harangues of most of our rhetoricians—they are empty, resembling bodies without substance. One would need only to put them into French, and no one would read them.

Little by little men of letters at last recovered from this

kind of mania. It seems that we owe their change, at least in part, to the patronage of the great, who are quite happy to be learned on the condition that they can become so without trouble, and who wish to be able to judge a work of intelligence without hard study, in exchange for the benefits they promise to the author or for the friendship with which they think they honor him. Men began to understand that beauty lost none of its advantages for being expressed in the common tongue, that it even gained by becoming more accessible to the generality of men, and that there was no merit in saying common or ridiculous things in any language whatever, especially in those languages which of necessity were spoken the worst. Therefore, men of letters turned their thoughts to perfecting the vulgar tongues. They tried first to say in these languages what the ancients had said in theirs. But, as a result of that prejudice which had been so hard to uproot, they began by distorting the French tongue instead of enriching it. Ronsard created a barbarous jargon, bristling with Greek and Latin; but fortunately he made is so unrecognizable that it became ridiculous. Men soon saw that it was the beauties and not the words of ancient languages which should be translated into our own. Guided and perfected by good taste, it rather quickly acquired an infinite number of pleasing and happy expressions. Finally, men no longer limited themselves to copying the Romans and the Greeks, or even to imitating them; they tried to surpass them, if possible, and to think according to their own intelligence. Thus, the imagination of the moderns was reborn little by little from that of the ancients; and all the masterpieces of the past century were seen to burst forth almost simultaneously—masterpieces in eloquence, in history, in poetry, and in the different literary genres.

Malherbe, nurtured by the reading of the excellent poets of antiquity and taking nature as a model as they had done, was the first to enhance our poetry with a harmony and beauty which it had never known before. Balzac, who is today held too much in contempt, gave nobility and balance to our prose.[8] The writers of Port-Royal continued what Balzac had begun.[9] They added that precision, that happy choice of terms, and that purity which make most of their works seem almost modern even now and which set them apart from a great many antiquated books written at the same time. Corneille [1606–1684], after having for several years made concessions to poor taste in his career as a dramatist, finally freed himself of it; by the strength of his genius more than by reading he discovered the laws of the theater and set them forth in his admirable Discourses on tragedy, in his Reflections upon each of his plays, but principally in his plays themselves. Racine [1639–1699], opening another route, brought to the stage a passion that the ancients had hardly known. Unfolding the motives of the human heart, he joined a consistent elegance and truth

---

[8] The national character of the *Discourse* is evident in d'Alembert's treatment of the Renaissance almost exclusively in terms of the history of French literature and art. His prejudice against the poets of the French Renaissance was shared by the majority of his enlightened contemporaries. It had originated in the seventeenth century and was raised to the level of dogma by the famous *Art poétique* of Boileau-Despréaux (1636–1711), whose work d'Alembert later called "'The code of good taste in our language, as that of Horace is in Latin" (*Œuvres*, II, 355, "*Éloge de Despréaux*"). Ronsard (1524–1585), the greatest of the French poets of his century, was dismissed as a pedant. Like Boileau, d'Alembert thought that French poetry had to await the coming of Malherbe (1555–1628), the court poet of Henry IV. Prose was not "perfected" until the works of Guez de Balzac (1597–1654). In the first half of the nineteenth century, French critics and poets rehabilitated many of the poets of the sixteenth century, and they have been read and enjoyed ever since, while Malherbe and Balzac are no longer in such high favor.

[9] The abbey of Port-Royal served as the center of the Jansenists, a group of French Catholic laymen and clergymen who resembled Calvinists in certain of their theological doctrines and in their austerity, and who were seriously persecuted by Louis XIV and officially condemned by the Pope. The prose of certain Jansenists such as Arnauld and Nicole was highly esteemed, and that of Pascal (1623–1662), the great scientist and lay theologian who was associated with the Jansenist cause for many years, is undoubtedly one of the glories of the French language.

with certain traits of the sublime. In his *Art poétique,* Despréaux made himself the equal of Horace at the same time that he modeled his work after him.[10] Molière [1622-1673] left ancient comedy far behind him in the subtle portrayal of the absurdities and the mores of his time. La Fontaine [1621-1695] almost caused Aesop and Phaedrus to be forgotten, and Bossuet succeeded in putting himself in the same rank with Demosthenes.[11]

The Fine Arts are so closely united with belles-lettres that the taste which cultivates one impels men to perfect the other. During the same time that our literature was being enriched by so many fine works, Poussin produced his paintings and Puget created his statues; Le Sueur painted the cloister of the Chartreux, and Le Brun the battles of Alexander; finally [12] Lully, creator of song appropriate for our language, gave immortality to the poems of Quinault through his music, which in turn was made immortal by them.[13]

[10] See n. 8, p. 67.

[11] Bishop Bossuet (1627-1704), the foremost defender of the Catholic faith in the late seventeenth century, also ranked among the literary masters of the classical age for the magnificent prose of his sermons and his historical writing.

[12] Later editions: "finally Quinault, creator of a new kind of work, assured himself immortality by his lyrical poems; and Lully. . . ."

[13] Note that nearly all of the artists whom d'Alembert eulogizes here were closely associated with the royal court, center of the finest expression of seventeenth-century French classical forms, similarly admired by Voltaire in his *Age of Louis XIV*. Nicolas Poussin (1594-1665) was famed for his historical scenes and landscapes painted in the classical manner. Pierre Puget (1622-1694) applied the same style to sculpture and architecture. Eustache Le Sueur (1616-1655) painted a celebrated series of scenes from the life of Saint Bruno in the cloister of the Paris Chartreux monastery and other works on religious themes. Louis XIV commissioned Charles Le Brun (1619-1690) to paint a series of scenes from the life of Alexander the Great at the palace of Fontainebleau and also employed him for eighteen years to decorate the palace of Versailles. The poet and playwright Philippe Quinault (1635-1688) wrote a number of librettos for the operas of Jean-Baptiste Lully (1633?-1687), court musician of Louis XIV, who was considered the founder of French opera and the finest of all French composers until Rameau.

It must be admitted, however, that the renaissance of painting and sculpture was much more rapid than that of poetry and music, and the reason is not difficult to understand. After men began to study ancient works of every kind, the rather large number of masterworks of antiquity that had escaped from superstition and barbarism immediately caught the attention of enlightened artists. The only way to imitate sculptors like Praxiteles and Phidias [14] was to do exactly as they did, and talent needed only to be instructed by a discerning eye. Thus, it did not take Raphael and Michelangelo long to bring their art to a point of perfection which has not yet been surpassed. Since the object of painting and of sculpture falls more within the domain of the senses, these arts in general could not fail to precede poetry. The senses must have been more promptly affected by the palpable and sensible beauties of ancient statues than was the imagination in perceiving the fugitive and intellectual beauties of the ancient writers. Indeed, even when men's imagination began to perceive these beauties, their manner of imitating them, this servile copying in a foreign tongue, was defective and would almost certainly have impeded the progress of the imagination. Imagine for a moment our painters and sculptors deprived of the advantage they enjoyed in using the same raw materials as the ancients. If they, like our literary men, had wasted their time seeking out and approximating that material poorly instead of using another for the purpose of imitating the works they admired, their progress would doubtless have been much less rapid, and they would still not have found true marble.[15]

As for music, it could achieve a certain degree of perfection only much later, because it is an art which the moderns have been obliged to create. Time destroyed all the models which the ancients may have left us in this genre. Their writers, at

[14] Phidias was the foremost sculptor of Greece in the fifth century B.C., and Praxiteles in the fourth century.
[15] That is, while a Renaissance sculptor could, like the ancients, work in marble, the Renaissance writer had to transpose the ancient forms into a new "material"—the modern vernacular.

least those who remain for us, have conveyed only the most obscure knowledge on this subject, or only stories that are more likely to astound us than to instruct. Several of our scholars, perhaps activated by a sort of proprietary love, have alleged that we have carried this art much further than the Greeks. Because of the lack of remains from antiquity, it is as difficult to support as to destroy this claim, and the true or supposed prodigies of ancient music are only feeble proof against it. Perhaps we could conjecture with some probability that ancient music was completely different from ours and that if the ancient was superior in melody, harmony gives certain advantages to the modern.

We would be unjust if we did not acknowledge what we owe to Italy in the matters we have just been discussing. It is from her that we have received the sciences which have since flourished so abundantly in all Europe. It is to her above all that we owe the fine arts and good taste, of which she has furnished us a large number of inimitable models.

While the arts and belles-lettres were held in honor, philosophy was far from making the same progress, at least in each nation taken as a whole; it reappeared only much later. It is not that to excel in belles-lettres is essentially easier than in philosophy; superiority in all genres is equally difficult to attain. But the reading of the ancients was to contribute more promptly to the advancement of belles-lettres and of good taste than to the progress of the natural sciences. Literary beauties do not have to be viewed for a long time in order to be felt; and as men feel before they think, for the same reason they judge their sensations before judging their thoughts. Moreover, as philosophers the ancients did not approach the perfection they achieved as writers [of literature], and indeed, though in the sequence of our ideas the first operations of reason precede the first efforts of the imagination, the imagination moves much faster than reason once it has made its first steps. Imagination has the advantage of operating on objects which it produces itself, whereas reason all too often exhausts itself in fruitless investigations, since perforce it is limited to the ideas which lie before it, and forced to check itself at each instant. The universe and reflections [of the mind] are the first book of the true philosophers, and the ancients doubtless had studied it; thus men should have done what the ancients did. That study could not be replaced by a study of their works, most of which had been destroyed. The small number [that survived], mutilated by time, could give us only very uncertain and much corrupted ideas on so vast a matter.

Scholasticism, which constitutes the whole of so-called science of the centuries of ignorance, still was prejudicial to the progress of true philosophy in that first century of enlightenment. Since time immemorial, so to speak, men had been persuaded that they possessed the doctrine of Aristotle in all its purity, [even though it had been] commented on by the Arabs and corrupted by thousands of absurd or childish additions. So great was their respect for the ancients that they did not even think of ascertaining whether that barbarous philosophy was really the philosophy of this great man. It is in this fashion that many peoples, born and confirmed by education in their errors, believe in all sincerity that they are on the road to truth, because it has never even occurred to them to form the least doubt about it. And so, in the time when several writers were rivals of the Greek orators and poets, marching in the same rank as their models and perhaps even surpassing them, Greek philosophy, although very imperfect, was not even well known.

This multitude of prejudices, which was partially supported by blind admiration for antiquity, seemed to be strengthened further by the abuse which a few powerful theologians dared to make of the submission of peoples. I say a few, for I am far from extending an accusation that concerns only some of its members to an entire respectable and highly enlightened body.[16] Poets were permitted to sing of the divinities of paganism, because men were rightly persuaded that using the names of these divinities could be no more than a game from which there was nothing to fear. If, on the one hand, the religion of

---

[16] This sentence is left out of the 1764 edition.

the ancients, thinking of every object as being alive, opened a broad scope to the imagination of literary men, on the other hand its principles were so absurd that no one needed to worry about a revival of Jupiter or Pluto by some sect of innovators. But some people did fear, or appeared to fear, the blows which blind reason might deliver against Christianity. How did they fail to see that such a feeble attack was not of the slightest danger to them? Christianity was sent to men from heaven, and the promises of God himself assured forever that righteous and ancient veneration which people manifested towards it. Moreover, however absurd a religion might be (a reproach which only impiety can make of ours), it is never the philosophers who destroy it. Even when they teach truth, they are satisfied to demonstrate it without forcing recognition from anyone. Such a power belongs only to the Omnipotent Being. It is the inspired men who enlighten the people and the enthusiasts who lead them astray. The bridle that we are obliged to impose upon the license of the latter should in no way harm that liberty which is so necessary to true philosophy and from which religion can draw the greatest advantages. If Christianity brings to philosophy the enlightenment that it lacks and if grace alone can force the incredulous to submit, it is reserved for philosophy to reduce them to silence. To assure the triumph of faith, the theologians of whom we speak needed only to employ those weapons which were supposed to constitute a threat to it.

But some among these men had much more compelling reasons to oppose the advance of philosophy. Falsely persuaded that the faith of peoples becomes firmer as the different objects upon which it is exercised become more numerous, they were not content to require a legitimate submission to our mysteries. They tried to elevate their individual opinions into dogmas. And it was these opinions themselves, far more than the dogmas, which they wanted to make secure. They would by this means have inflicted the most terrible blow upon religion, had religion been the work of men. For there was a danger that once these opinions were recognized as false,

PRELIMINARY DISCOURSE 73

the common people (who have no discernment) might treat the truths of religion and the false opinions with which some had wished to confound them in the same way.

Other theologians who were of better faith, but equally dangerous, joined with the first for different motives. Although religion is intended uniquely to regulate our mode of life and our faith, they believed it was to enlighten us also on the system of the world—in short, on matters which the All-Powerful has expressly left to our own disputations. They did not make the reflection that the sacred books and the works of the Fathers, which were created to teach the common people as well as the philosophers the requirements of practice and belief, would have spoken only the language of the common people when it came to indifferent questions. However, theological despotism or prejudice won out. A tribunal whose name still cannot be spoken without fear in France became powerful in the south of Europe, in the Indies, and the New World. Faith in no way ordained belief in it, nor charity the approval of it.[17] It condemned a celebrated astronomer for having maintained that the earth moved and declared him a heretic, almost in the way that Pope Zachary had, some centuries before, condemned a bishop for not having thought as St. Augustine did concerning the antipodes, and for having guessed their existence six hundred years before Christopher Columbus discovered them.[18] It was thus that the abuse of the

[17] Later editions add the phrase: "rather, religion reproves it, although this tribunal is occupied by its ministers."

[18] The tribunal to which d'Alembert refers is, of course, the Inquisition; Galileo (1564–1642) was the astronomer condemned by it. The Jesuit *Journal de Trévoux* (October, 1751), in its critical review of the first volume of the *Encyclopedia*, took issue with d'Alembert's reference to Pope Zachary (d. 752), among other things pointing out that it was not a bishop but a priest who was condemned and that he had been accused of asserting that there were races of men who were not descended from Adam. D'Alembert gave over a large section of the article "Antipodes," *Encyclopédie*, I, 512–14, to a more sharply anticlerical discussion of this question, in which he acidly assailed an author in the *Journal de Trévoux* of 1708, who tried to justify this action of the Pope. In the *errata* of Volume II, page iv, he points out that the man in ques-

spiritual authority, conjoined with the temporal, forced reason to silence; and they were not far from forbidding the human race to think.

While poorly instructed or badly intentioned adversaries made open war on it, philosophy sought refuge, so to speak, in the works of a few great men. They had not the dangerous ambition of removing the blindfolds from their contemporaries' eyes; yet silently in the shadows they prepared from afar the light which gradually, by imperceptible degrees, would illuminate the world.

The immortal Chancellor of England, Francis Bacon [1561–1626], ought to be placed at the head of these illustrious personages. His works, so justly esteemed (and more esteemed, indeed, than they are known), merit our reading even more than our praises.[19] One would be tempted to regard him as the greatest, the most universal, and the most eloquent of the philosophers, considering his sound and broad views, the multitude of objects to which his mind turned itself, and the boldness of his style, which everywhere joined the most sublime images with the most rigorous precision. Born in the depths of the most profound night, Bacon was aware that philosophy did not yet exist, although many men doubtless flattered themselves that they excelled in it (for the cruder a century is, the more it believes itself to be educated in all that can be known).

---

tion may have been only a priest rather than a bishop, and he made a rather contemptuous reference to this criticism and others by the *Journal de Trévoux* in the preface to the second edition of the *Discourse*, in *Œuvres*, I, 14. Lynn Thorndike discusses this problem in detail in his article "*L'Encyclopédie* and the History of Science," *Isis*, VI (1924), 369.

[19] According to F. M. Grimm, (*Correspondance littéraire*, III, 116) Diderot was the first serious student of Bacon among the group of encyclopedists, and Diderot himself claims in the article "Encyclopédie," V, 647, to have taught his fellow citizens to esteem and read Bacon. It is apparent that Condillac had not read Bacon until he was writing his *Essai sur l'origine des connoissances humaines* (1746), in *Œuvres philosophiques*, I, 115, and d'Alembert gives no indication that he had made a study of Bacon before he wrote the *Discourse*. D'Alembert's historical and biographical judgments on Bacon are close to those in Voltaire's *Lettres philosophiques*, Letter XII. See Voltaire, *Œuvres*, XXII, 116 ff.

Therefore, he began by considering generally the various objects of all the natural sciences. He divided these sciences into different branches, of which he made the most exact enumeration that was possible for him. He examined what was already known concerning each of these objects and made the immense catalogue of what remained to be discovered. This is the aim of his admirable book *The Advancement of Learning*. In his *Novum Organum*, he perfects the views that he had presented in the first book, carries them further, and makes known the necessity of experimental physics, of which no one was yet aware. Hostile to systems, he conceives of philosophy as being only that part of our knowledge which should contribute to making us better or happier, thus apparently confining it within the limits of the science of useful things, and everywhere he recommends the study of Nature. His other writings were produced on the same pattern. Everything, even their titles, proclaims the man of genius, the mind that sees things in the large view. He collects facts, he compares experiments and points out a large number to be made; he invites scholars to study and perfect the arts, which he regards as the most exalted and most essential part of human science; he sets forth with a noble simplicity his *Conjectures and Thoughts* on the different objects worthy of men's interest;[20] and he would have been able to say, like that old man in Terence, that nothing which touches humanity was alien to him. Natural science, ethics, politics, economics, all seem to have been within the competence of that brilliant and profound mind. And we do not know which we ought to admire more, the riches he lavishes upon all the subjects he treats or the dignity with which he speaks of them. His writings can best be compared to those of Hippocrates on medicine; and they would be no less admired, nor less read, if the culture of the mind were as dear to mankind as the conservation of health. But in every area only the works of those who head a school of disciples

[20] This is probably a reference to Bacon's *Cogitata et visa* (published posthumously). It might, however, refer to Bacon's *Essays*, some of which d'Alembert translated into French (*Œuvres*, IV, 227 ff.).

make a brilliant impression. Bacon was not of that number, and the form of his philosophy prevented it; it was too wise to astonish anyone. Scholasticism, which continued to dominate, could not be overthrown except by bold and new opinions. And apparently circumstances are not such that a philosopher who is content to say to men: "Here is the little that you have learned, there is what remains for you to find," is destined to cause much stir among his contemporaries. If we did not know with what discretion, and with what superstition almost, one ought to judge a genius so sublime, we might even dare reproach Chancellor Bacon for having perhaps been too timid. He asserted that the scholastics had enervated science by their petty questions, and that the mind ought to sacrifice the study of general beings for that of individual objects; nonetheless, he seems to have shown a little too much caution or deference to the dominant taste of his century in his frequent use of the terms of the scholastics, sometimes even of scholastic principles, and in the use of divisions and subdivisions, fashionable in his time. After having burst so many irons, this great man was still held by certain chains which he could not, or dared not, break.

We declare here that we owe principally to Chancellor Bacon the encyclopedic tree of which we have already spoken at length and which will be found at the end of this Discourse. We have stated this in several places in the *Prospectus*, we come back to it again, and we will not lose any opportunity to repeat it. However, we have not believed it necessary to follow on every point the great man whom we acknowledge here to be our master. If we have not put reason after imagination as he did, it is because we have followed the metaphysical order of the operations of the mind in the encyclopedic system rather than the historical order of its progress since the renaissance of letters. The illustrious Chancellor of England perhaps had this latter order in mind to some extent when he made what he called "the census" and the enumeration of the parts of human knowledge. Moreover, since the plan of Bacon is different from ours, and since the sciences have subsequently made great progress, it should not be surprising that we have sometimes taken a different route.

Besides the changes that we have already explained we have made in the sequence of the general distribution of knowledge, we have in certain respects pushed the subdivisions further, especially in the part on mathematics and on particular physics.[21] On the other hand, we have abstained from extending as far as he does the subdivision of certain sciences, which he follows to the last twigs of their branches. These, which ought properly to enter into the body of our Encyclopedia, would only have enlarged the "general system" rather uselessly in our opinion. The encyclopedic tree of the English philosopher will be found immediately following ours. That is the shortest and easiest way to distinguish between what comes from us and what we have borrowed from him.[22]

Chancellor Bacon was followed by the illustrious Descartes [1596–1650]. That exceptional man, whose fortune has varied so much in less than a century, possessed all the qualities necessary for changing the face of philosophy: a strong imagination, a most logical mind, knowledge drawn from himself more than from books, great courage in battling the most generally accepted prejudices, and no form of dependence which forced him to spare them. Consequently, he experienced even in his own life what ordinarily happens to any man who has too marked an ascendancy over others. He had few enthusiasts and many enemies. Whether because he knew his country or only because he distrusted it, he took refuge in an entirely free land in order to meditate with less possibility of disturbance. Though he was much less concerned with attracting disciples than with deserving them, persecution went to search him out

---

[21] "Particular physics" is defined in the "Detailed Explanation of the System of Human Knowledge." (See pp. 152 and 154 ff.)

[22] D'Alembert takes special pains to point out what the *Encyclopedia* borrowed from Bacon and what was added, in order to refute the implied charge of plagiarism in the *Journal de Trévoux*, January, 1751, p. 188. Diderot's outline of the Baconian systematization of knowledge is included at the end of the *Discourse*.

in his retreat, and the secluded life he led could not protect him from it. In spite of all the sagacity that he had employed to prove God's existence, he was accused of denying it by some officials who perhaps did not believe in it themselves. Tormented and slandered by foreigners, and rather poorly received by his compatriots, he went to die in Sweden, doubtless far from anticipating the brilliant success that his opinions would one day have.

One can view Descartes as a geometer or as a philosopher. Mathematics, which he seems to have considered lightly, nevertheless today constitutes the most solid and the least contested part of his glory. Algebra, which had been virtually created by the Italians and prodigiously augmented by our illustrious Viète, received new acquisitions in the hands of Descartes.[23] One of the most considerable is his method of indeterminates, a very ingenious and very subtle artifice, which we have since been able to apply to a large number of investigations. But above all what immortalized the name of this great man is the application he was able to make of algebra to geometry, one of the grandest and most fortunate ideas that the human mind has ever had. It will always be the key to the most profound investigations, not only in sublime geometry, but also in all the physico-mathematical sciences.

As a philosopher he was perhaps equally great, but he was not so fortunate. Geometry, which by the nature of its object always gains without losing ground, could not fail to make a progress that was most sensible and apparent for everyone when plied by so great a genius. Philosophy found itself in quite a different state. There everything remained to be done, and what do not the first steps in any branch of knowledge cost? One is excused from making larger steps by the merit of taking any at all. If Descartes, who opened the way for us, did not progress as far along it as his sectaries believe, nevertheless the sciences are far more indebted to him than his adversaries will allow; his method alone would have sufficed to render him

[23] François Viète (1540–1603) was a pioneer French algebraist and geometer who introduced the use of letters for quantities in algebra.

immortal. His *Dioptrics* is the greatest and the most excellent application that has yet been made of geometry to physics. In a word, we see his inventive genius shining forth everywhere, even in those works which are least read now. If one judges impartially those vortices which today seem almost ridiculous, it will be agreed, I daresay, that at that time nothing better could be imagined.[24] The astronomical observations which served to destroy them were still imperfect or hardly established. Nothing was more natural than to postulate a fluid which carried the planets. Only a long sequence of phenomena, of reasonings, and of calculations, and consequently a long sequence of years, could cause such an attractive theory to be renounced. Moreover, it had the singular advantage of explaining gravitation of bodies by the centrifugal force of the vortex itself, and I am not afraid to assert that this explanation of weight is one of the finest and most ingenious hypotheses that philosophy has ever imagined. Thus, physicists had to be carried forward almost in spite of themselves by the theory of central forces and by experiments made much later before they would abandon it. Let us recognize, therefore, that Descartes, who was forced to create a completely new physics, could not have created it better; that it was necessary, so to speak, to pass by way of the vortices in order to arrive at the true system of the world; and that if he was mistaken concerning the laws of movement, he was the first, at least, to see that they must exist.[25]

His metaphysics, as ingenious and new as his physics, suf-

[24] Descartes employed the vortex theory to account for the circular movement of planets, which Newton later explained by gravity. The physics of Descartes postulated that whenever a physical object moves, something must move into its place, since there is no vacuum in the physical world. The entire world of matter is characterized from one end to the other by a circular or vortical motion of bodies, there being innumerable such vortices in the universe, including the great one of the planets around the sun.

[25] The prefaces to his earliest scientific works show how carefully d'Alembert had studied the work of Descartes and the great respect he had for him as a physical scientist and mathematician.

fered virtually the same fate. And it too can be justified by more or less the same reasons; for such is the fortune of that great man today, that after having had innumerable disciples, he is reduced to a handful of apologists. No doubt he was mistaken in admitting the existence of innate ideas. But had he retained that single truth taught by the Aristotelians concerning the origin of ideas through the senses, perhaps it would have been more difficult to uproot the errors that debased this truth by being alloyed with it. Descartes dared at least to show intelligent minds how to throw off the yoke of scholasticism, of opinion, of authority—in a word, of prejudices and barbarism. And by that revolt whose fruits we are reaping today, he rendered a service to philosophy perhaps more difficult to perform [26] than all those contributed thereafter by his illustrious successors. He can be thought of as a leader of conspirators who, before anyone else, had the courage to arise against a despotic and arbitrary power and who, in preparing a resounding revolution, laid the foundations of a more just and happier government, which he himself was not able to see established. If he concluded by believing he could explain everything, he at least began by doubting everything, and the arms which we use to combat him belong to him no less because we turn them against him. Moreover, when absurd opinions are of long duration, one is sometimes forced to replace them by other errors, if one cannot do better, in order to disabuse the human race. The uncertainty and the vanity of the mind are such that it always needs an opinion to which to cleave. It is a child to whom one must offer a toy in order to take a dangerous weapon away from him. He will quit this toy by himself when he reaches the age of reason. In thus deceiving the philosophers, or those who believe themselves such, one teaches them at least to distrust their intelligence, and that frame of mind is the first step toward truth. Thus was Descartes persecuted in his own lifetime, as if he had come to bring truth to men.

Newton [1642–1727], whose way had been prepared by Huy-

[26] In the 1764 edition d'Alembert substitutes here: "more essential than all those contributed. . . ."

ghens, appeared at last, and gave philosophy a form which apparently it is to keep. That great genius saw that it was time to banish conjectures and vague hypotheses from physics, or at least to present them only for what they were worth, and that this science was uniquely susceptible to the experiments of geometry.[27] It was perhaps with this aim that he began by inventing calculus and the method of series, whose applications are so extensive in geometry itself and still more so in explaining the complicated effects that one observes in Nature, where everything seems to take place by various kinds of infinite progressions. The experiments on weight and the observations of Kepler led the English philosopher to discover the force which holds the planets in their orbits. Simultaneously he showed how to distinguish the causes of their movements and how to calculate them with a precision such as one might reasonably expect only after several centuries of labor. Creator of an entirely new optics, he made the properties of light known to men by breaking it up into its constituent parts. Anything we could add to the praise of the great philosopher would fall far short of the universal testimonial that is given today to his almost innumerable discoveries and to his genius, which was at the same time far-reaching, exact, and profound. He has doubtless deserved all the recognition that has been given him for enriching philosophy with a large quantity of real assets. But perhaps he has done more by teaching philosophy to be judicious and to restrict within reasonable limits the sort of audacity which Descartes had been forced by circumstances to bestow upon it. His Theory of the World (for I do not mean his System) is today so generally accepted that men are beginning to dispute the author's claim to the honor of inventing it (because at the beginning great men are accused of being mistaken, and at the end they are treated as plagiarists). To the people who find everything in the works of the ancients I leave the pleasure of discovering gravitation of planets in those works, even if it is not there. But even supposing that the Greeks had had the idea of it, what was only a rash and ro-

---

[27] Later editions: "to experiments and geometry."

mantic system with them became a demonstration in the hands of Newton. That demonstration, which belongs exclusively to him, constitutes the real merit of his discovery. Without such support, the theory of attraction would only be an hypothesis like so many others. If today some celebrated writer takes it into his head to predict without any proof that one day we will succeed in making gold, would our descendants be justified in trying on this pretext to deprive a chemist who actually achieved it of the glory of the great work? And does the invention of telescopes belong any less to its authors, even though some ancient might have thought it possible that we would one day extend the range of our vision?

Other scholars believe they are making a far better founded criticism of Newton by accusing him of reviving the "occult qualities" of the scholastics and ancient philosophers in physics.[28] But are the scholars of whom we are speaking quite sure that these two words were anything for the ancient philosophers but the modest expression of their ignorance? They were empty of sense among the scholastics, though intended to designate a being of which they thought they could conceive. Newton, who had studied Nature, did not flatter himself that he knew more than the ancients concerning the first cause which produces phenomena. But he refrained from employing the same language, in order not to offend some contemporaries who would have attached an idea to it other than what he intended. He contented himself with proving that the vortices of Descartes could not reasonably explain the movement of the planets, that phenomena and the laws of mechanics unite in overthrowing the vortices, that there is a force through which the planets tend toward one another, whose principle is entirely unknown to us. He did not reject impulsion; he limited

---

[28] Joseph Saurin (1659–1737), a French Cartesian mathematician and physicist, asserted in the first decade of the eighteenth century that the universal "attraction" of Newton, which could be only measured and not explained, was nothing more than a new version of the scholastic "occult qualities." See Fontenelle's "Éloge de M. de Saurin," Œuvres (Amsterdam, 1764), VI, 338; Voltaire's Lettres philosophiques, in Œuvres, XXII, 138; and d'Alembert's article "Attraction," Encyclopédie, I, 854.

himself to asking that it be employed more felicitously than it had been up to that time to explain the movement of the planets. His desires have not yet been fulfilled and perhaps will not be for a long time. After all, by giving us grounds to think that matter may have properties which we did not suspect and by disabusing us of our ridiculous confidence that we know everything, what harm could he have done to philosophy?

It appears that Newton had not entirely neglected metaphysics. He was too great a philosopher not to be aware that it constitutes the basis of our knowledge and that clear and exact notions about everything must be sought in metaphysics alone. Indeed, the works of this profound geometer make it apparent that he had succeeded in constructing such notions for himself concerning the principal objects that occupied him. However, he abstained almost totally from discussing his metaphysics in his best known writings, and we can hardly learn what he thought concerning the different objects of that discipline, except in the works of his followers. This may have been because he himself was somewhat dissatisfied with the progress he had made in metaphysics, or because he believed it difficult to give mankind sufficiently satisfactory and extensive enlightenment on a discipline too often uncertain and disputed. Or finally, it may have been because he feared that in the shadow of his authority people might abuse his metaphysics as they had abused Descartes', in order to support dangerous or erroneous opinions. Therefore, since he has not caused any revolution here, we will abstain from considering him from the standpoint of this subject.

Locke [1632–1704] undertook and successfully carried through what Newton had not dared to do, or perhaps would have found impossible. It can be said that he created metaphysics, almost as Newton had created physics. He understood that the abstractions and ridiculous questions which had been debated up to that time and which had seemed to constitute the substance of philosophy were the very part most necessary to proscribe. He sought the principal causes of our errors in those abstractions and in the abuse of signs, and that is where he found them. In order to know our soul, its ideas, and its affections, he did not study books, because they would only have instructed him badly; he was content with probing deeply into himself, and after having contemplated himself, so to speak, for a long time, he did nothing more in his treatise, *Essay Concerning Human Understanding* [1690], than to present mankind with the mirror in which he had looked at himself. In a word, he reduced metaphysics to what it really ought to be: the experimental physics of the soul—a very different kind of physics from that of bodies, not only in its object, but in its way of viewing that object. In the latter study we can, and often do, discover unknown phenomena. In the former, facts as ancient as the world exist equally in all men; so much the worse for whoever believes he is seeing something new. Reasonable metaphysics can only consist, as does experimental physics, in the careful assembling of all these facts, in reducing them to a corpus of information, in explaining some by others, and in distinguishing those which ought to hold the first rank and serve as the foundation. In brief, the principles of metaphysics, which are as simple as axioms, are the same for the philosophers as for the general run of people. But the meager progress that this science has made for such a long time shows how rarely these principles are applied felicitously, whether because of the difficulty that surrounds such an application, or perhaps also because of the natural temptations that prevent us from holding ourselves within bounds when we engage in metaphysical speculations. Nevertheless, the title of metaphysician, and even of great metaphysician, is still rather common in our century, for we love to squander everything. But how few persons are really worthy of that name! How many are there who earn it only by the unfortunate talent of obscuring clear ideas with a great deal of subtlety and of preferring their own extraordinary notions to true ones which are always simple? One should not be astonished, therefore, if the majority of those who are called metaphysicians have so little regard for one another. I have no doubt that this title will soon become an insult for our men of intelligence, as the name "sophist," degraded by those who bore it in Greece, was rejected by true philosophers, even though it means "a sage."

We may conclude from all this history that England is indebted to us for the origins of that philosophy which we have since received back from her. It is perhaps a greater distance from substantial forms to vortices than from vortices to universal gravitation; just as perhaps a greater interval exists between pure algebra and the idea of applying it to geometry than between the small triangle of Barrow and differential calculus.[29]

Such are the principal geniuses that the human mind ought to regard as its masters. Greece would have raised statues to them, even if she had been obliged to tear down those of a few conquerors in order to give them room.

The limits of the Preliminary Discourse prevent us from speaking of several illustrious philosophers who, without proposing views as great as those which we have just mentioned, have contributed much to the advancement of the sciences through their works and lifted, so to speak, a corner of the veil that concealed truth from us. Among these are: Galileo [1564–1642], whose astronomical discoveries contributed so much to geography, as did his theory of acceleration to mechanics; Harvey [1578–1657], who will be immortalized by his discovery of the circulation of blood; Huyghens, whom we have already named, and whose forceful and ingenious works have served geometry and physics so well;[30] Pascal, author of a treatise on the cycloid which ought to be regarded as a prodigy of wisdom and of penetration, and of a treatise on the equilibrium of liquids and the weight of air which has opened a new science to us—a sublime and universal genius whose talents could not

---

[29] Isaac Barrow (1630–1677) was Newton's professor, who brought mathematics to the verge of the discovery of calculus.

[30] Christian Huyghens (1629–1695), a Dutch physicist, mathematician, and astronomer, is remembered for his study of the pendulum. A contemporary of Newton and Leibniz and a "reformed" disciple of Descartes, he was one of the great scientists of his age. For a number of years he lived in Paris.

be too much mourned by philosophy, if religion had not profited from them; [31] Malebranche, who unraveled the errors of the senses so well, and who knew the errors of the imagination—as if he had not often been misled by his own; [32] Boyle [1627–1691], the father of experimental physics; and several others, among whom men like Vesalius, Sydenham, Boerhaave, and an infinite number of celebrated anatomists and physical scientists ought to be counted with distinction.[33]

Among these great men there is one whose philosophy, which is today both very well received and strongly opposed in the north of Europe, obliges us not to pass over him in silence. This is the illustrious Leibniz. If he had had nothing more than the glory of sharing, or even the suspicion of sharing, the invention of differential calculus with Newton, he would deserve honorable mention on that score alone.[34] But we wish to examine him principally for his metaphysics. Like Descartes, he seems to have recognized the inadequacy of all previous solutions to those lofty questions concerning the union of the body and the soul, Providence, and the nature of mat-

[31] Blaise Pascal (1623–1662), the brilliant French scientist, mathematician, and (later) theologian, is especially noted for his discoveries concerning atmospheric pressure, the equilibrium of fluids, and the mathematics of probability. See n. 9, p. 67.

[32] Malebranche (1638–1715) was a French Roman Catholic cleric who developed a vast metaphysical system derived from the philosophy of Descartes. Perhaps his most famous idea was that we "see all things in God," who alone makes it possible for a relationship to exist between our ideas and their objects, since mind and matter are completely diverse in nature. The *philosophes* found him guilty of engaging in the "spirit of systems."

[33] Vesalius (1514–1564) was the most renowned of the early modern anatomists, Sydenham (1624–1689) an English physician and pioneer modern clinician, and Boerhaave (1668–1738), a Dutch physician and professor of Leyden who in his day was almost as celebrated as Newton and Leibniz. He applied the Newtonian experimental method to medicine and chemistry.

[34] A most unpleasant dispute broke out between Leibniz and Newton and their respective partisans over which man had a right to claim the glory of inventing calculus. It is almost certain that each made his discovery independently.

ter. He seems even to have had the advantage of setting forth the difficulties which can be raised on these questions more forcefully than anyone else. But, less judicious than Locke and Newton, he was not contented with formulating doubts; he tried to dissipate them, and in that respect he was perhaps no more fortunate than Descartes. His principle of *sufficient reason*, excellent and very true in itself, does not seem very useful to beings so poorly enlightened as we regarding the primary reasons of all things. His *monads* at most prove that he understood better than anyone else our incapacity to form a clear idea of matter; but they do not appear to be such as to give us that clear idea. His *pre-established harmony* seems only to add a further difficulty to the opinion of Descartes concerning the union of the body and the soul.[35] Finally, his system of *Optimism* is perhaps dangerous because of the alleged advantage that it has of explaining everything.[36]

We will conclude with an observation which will not appear surprising to philosophers. The great men of whom we have just spoken changed the face of the sciences but little during their own lives. We have already seen why Bacon was not the head of a school of disciples. In addition to the reason which we have already given for this fact are two more. This great philosopher wrote several of his works in a seclusion to which his enemies had forced him, and the evil they had done to the statesman could not fail to harm the author as well. Moreover, uniquely occupied in being useful, he perhaps embraced too many matters for his contemporaries to

[35] Any competent history of philosophy will deal with the subjects of sufficient reason, monads, pre-established harmony. The encyclopedists were largely unresponsive to Leibniz because he was author of a philosophical "system" which went contrary to their factual and analytical leanings. He was, nonetheless, much admired for his awesome intellectual stature, and there are many references to him in d'Alembert's articles.

[36] Later editions included the following sentence: "This great man seems to have brought to metaphysics more sagacity than enlightenment, but however one thinks on this point, one cannot refuse him the admiration which he merits by the grandeur of his views of all kinds, the prodigious extent of his knowledge, and above all the philosophical spirit by which he was able to illuminate them."

allow themselves to be enlightened simultaneously on such a great number of objects. People hardly permit great geniuses to know so much; they are quite happy to learn something from them on a limited subject, but they do not want to be obliged to remodel all their ideas upon those of the great geniuses. Partly for this reason the works of Descartes underwent more persecution in France after his death than their author had suffered in Holland during his life; it was only with great difficulty that the schools finally dared admit a physics which they imagined to be contrary to that of Moses. It is true that Newton found less opposition among his contemporaries. This may have been because he first became known through his geometrical discoveries, which were of unquestioned propriety and reality, and these discoveries inclined people to admire and praise him more gradually and somewhat without constraint; or because his superior ability imposed silence on envy, or finally—and this seems more difficult to believe—because he was dealing with a nation that was less unjust than others. He had the singular advantage of seeing his philosophy generally accepted in England during his life and of having all his compatriots for partisans and admirers.[37] However, at that time it would have taken a good deal to make Europe likewise accept his works. Not only were they unknown in France, but scholastic philosophy was still dominant there when Newton had already overthrown Cartesian physics; the vortices were destroyed even before we considered adopting them. It took us as long to get over defending them as it did for us to accept them in the first place. It is necessary only to open our books in order to see with surprise that twenty years have not yet passed since we began to renounce Cartesianism in France. The first among us who

[37] Here is an illustration of a favorite device of the critics of the Old Régime in France, and especially of the encyclopedists. They employed the example of England's liberal virtues as a means of setting in relief the abuses in France. So great was the *philosophes'* admiration for the life and society of England in the eighteenth century that it came to be called *anglomanie*.

dared declare himself openly Newtonian was the author of the *Discours sur la figure des Astres* [1732], who combines a very extensive knowledge of geometry with the kind of philosophical mind not always found in conjunction with it, and also a talent for writing to which his geometrical knowledge certainly does no harm, as will be seen upon reading his works.[38] Maupertuis believed that one could be a good citizen without blindly adopting the physics of one's country; and we ought to be grateful to him for the courage he had to display in attacking that physics. In fact our nation, which is singularly eager for novelties in matters of taste, is on the contrary very much attached to ancient opinions in matters of science. Two such apparently contrary dispositions have their foundation in several causes, above all in that eagerness to enjoy, which seems to constitute our character. Things that have to do with feeling are not of a nature to be sought after for long, and they cease to be attractive when they are not presented all at once; but at the same time the ardor with which we apply ourselves to matters of feeling is soon exhausted. Disenchanted as soon as it is sated, the soul flies toward a new object, which it will likewise abandon. On the contrary, only through meditation does the mind achieve what it is after. For that very reason it tries to enjoy for as long as it has sought, especially when it is only a matter of an hypothetical and conjectural philosophy, which is much less toilsome than exact calculations and combinations. Physicists, who are attached to their theories with the same zeal and for the same motives as artisans to their practices on this point, resemble the common run of people

---

[38] Maupertuis (1698–1759) was the eminent and controversial French mathematician, astronomer, and Newtonian physicist selected by Frederick II of Prussia to be the president of his Royal Academy of Sciences, a choice that was commended by d'Alembert, who was once his protegé, in the article "Académie royale des sciences," *Encyclopédie*, I, 55. The Prussian king on a number of occasions offered the succession to this honor to his friend d'Alembert, both before and after the death of Maupertuis. D'Alembert cites the work of Maupertuis throughout the scientific articles of the *Encyclopedia*.

much more than they imagine. Let us always respect Descartes, but let us readily abandon opinions which he himself would have combatted a century later. Above all, let us not confound his cause with that of his sectaries. The genius that he manifested in seeking out a new, albeit false, route in the darkest night was unique with him. Those who have dared make the first steps into darkness have at least shown courage; but once it is light there is no longer any glory in losing one's way following those first footsteps. Of the few remaining partisans of his doctrine, he himself would have disavowed those who defend it purely out of slavish attachment to what they learned in their infancy or through some sort of national prejudice, the shame of philosophy. With such motives one can be the last of his partisans, but one would not have had the merit of being his first disciple. Rather, such a person would have been Descartes' adversary at a time when to be so would have been entirely unjust. In order to have the right to admire the errors of a great man, one must be able to recognize them when time has brought them out into broad daylight. Thus young men, whom we ordinarily regard as rather bad judges, are perhaps the best judges in philosophical and many other matters, when they are not deprived of enlightenment. Everything is equally new to them; thus their sole interest lies in making the right choice.

In fact, it is the young geometers in France as well as in foreign countries who have directed the fate of the two philosophies. The old one is proscribed to such an extent that its most zealous partisans no longer even dare mention the vortices with which they formerly stuffed their works. If Newtonianism were to be destroyed in our time by any cause whatsoever—the numerous partisans that it now has would doubtless play the same role that they have made others play. Such is the nature of minds; such are the results of self-esteem, which governs philosophers at least as much as other men, and of the opposition that all discoveries, both real and apparent, must meet.

Almost the same thing has been true of Locke as of Bacon,

Descartes, and Newton. Forgotten for a long time in favor of Rohault and Régis,[39] and still rather poorly known by the multitude, Locke is finally beginning to have some readers and a few partisans among us. It is thus that the efforts of illustrious persons, who are often too far above their time, are almost inevitably of no profit to their own centuries. It is reserved for following ages to receive the fruit of their enlightenment. Thus the restorers of the sciences almost never enjoy all the glory they merit. Far inferior men [40] wrest it away from them, because great men devote themselves to their own genius and mediocre people to that of their nation. It is true that the awareness of superiority, which is inevitably enjoyed by those who possess it, is sufficient to compensate for the lack of popular recognition. Superiority nourishes itself on its own substance; and often that repute for which people are so eager only serves to console mediocrity for the advantages that talent has over it. One may say, indeed, that Fame, as she tells all the news, relates hearsay more often than what she sees with her own eyes, and the poets who have portrayed her with a hundred mouths should also have given her a blindfold.

By the progress that it is making among us, philosophy, which constitutes the dominant taste of our century, seems to be trying to make up for the time that it has lost and to avenge itself for the sort of contempt our fathers showed for it. Today this disdain has fallen on erudition, and is no more just for having changed its object. Men imagine that we have [already] drawn everything worth knowing from the

---

[39] Jacques Rohault (1620–1675) and his disciple, Pierre Sylvain Régis (1632–1707) were the chief French interpreters and militant proponents of the philosophy and scientific theories of Descartes in the late seventeenth century. The conquest of French intellectual circles at the end of the century by the formerly proscribed Cartesianism owed much to these two men. See Aram Vartanian, *Diderot and Descartes* (Princeton, 1953), pp. 25, 54, 219–20. French philosophy and Cartesianism seemed so thoroughly identical at the beginning of the century that, as d'Alembert points out, resistance to foreign philosophies became, for some, almost a matter of national pride.

[40] 1764 edition: "minds."

works of the ancients, and on this basis they would willingly spare those who still wish to consult them the trouble. It seems that antiquity is regarded as an oracle that it is useless to consult because it has said everything it is going to say. Nowadays men have hardly more regard for the restitution of a lost passage than for the discovery of a small subdivision of a vein in the human body. But just as it would be ridiculous to believe that there is no more to be discovered in anatomy because the anatomists sometimes give themselves over to apparently useless researches (which are often useful by their consequences), it would be no less absurd to try to forbid erudition because of the rather unimportant researches our scholars may engage in. It would be ignorant and presumptuous to believe that everything is known concerning any subject whatsoever, and that we no longer have any advantage to draw from the study and reading of the ancients.

The practice today of writing everything in the vulgar tongue has doubtless contributed to strengthening this prejudice, and perhaps is more pernicious than the prejudice itself. Since our language was spread throughout all Europe, we decided that it was time to substitute it for Latin, which had been the language of our scholars since the renaissance of letters. I acknowledge that it is much more excusable for a philosopher to write in French than for a Frenchman to make Latin verses. I would be willing even to agree that this practice has contributed to making enlightenment more general, if indeed broadening the outer surface also broadens the mind within. However, an inconvenience that we certainly ought to have foreseen results from it. The scholars of other nations for whom we have set the example have rightly thought that they would write still better in their own language than in ours. Thus England has imitated us. Latin, which seemed to have taken refuge in Germany, is gradually losing ground there. I have no doubt that Germany will soon be followed by the Swedes, the Danes, and the Russians. Thus, before the end of the eighteenth century, a philosopher who would like to educate himself thoroughly concerning the discoveries of his predecessors will be required to burden his memory with seven or eight different languages. And after having consumed the most precious time of his life in learning them, he will die before beginning to educate himself. The use of the Latin language, which we have shown to be ridiculous in matters of taste, is of the greatest service in works of philosophy, whose merit is entirely determined by clarity and precision, and which urgently require a universal and conventional language. It is therefore to be hoped that this usage will be reestablished, yet we have no grounds to hope for it: the abuse of which we make bold to complain is too favorable to vanity and sloth for one to hope to uproot it. Philosophers, like other writers, want to be read, and above all by their own people. If they used a less familiar language, they would have fewer voices to praise them and a multitude could not boast of understanding them. It is true that with fewer admirers they would have better judges, but such an advantage affects them little, because fame has more to do with the number of those who give it than with their merit.

In recompense (for one must not overdo anything), our books of science appear to have acquired even the sort of advantage which seemed to belong specifically to works of belles-lettres. A respectable writer, whom our century still has the good fortune to possess and whose different productions I would praise here if I were not limiting myself to viewing him as a philosopher, has taught scholars to throw off the yoke of pedantry.[41] Superior in the art of elucidating the most abstract ideas, he has been able through the utmost method, precision, and clarity, to bring them down within the compass

[41] D'Alembert refers here to the magnificent Fontenelle (1657–1757), one of the creators of the Enlightenment and a favorite model for its writers, whose hundred-year life spanned the age of Louis XIV and a good part of the Enlightenment. He was a master at stating scientific and philosophical ideas in a comprehensible way, from whom the encyclopedists drew heavily for their articles, and, as secretary of the Académie des Sciences, he composed a celebrated collection of eulogies for its distinguished scientists. D'Alembert later performed a similar secretarial function at the Académie Française.

of minds which one would have believed least capable of grasping them. To philosophy he even dared lend the ornaments which seemed most foreign to it and which seemed to be most rigorously proscribed. And this boldness has been justified by the most general and flattering success. But like all original writers, he has left those who believed themselves able to imitate him far behind.

The author of *l'Histoire naturelle* has followed an entirely different route.[42] Rivaling Plato and Lucretius, he has lavished upon his work, whose reputation grows from day to day, that nobility and elevation of style which are so suited to philosophical matters and which, in the writings of the wise man, must be the picture of his soul.

But while intending to please, philosophy seems not to have forgotten that it is designed principally to instruct. For that reason the taste for systems—more suited to flatter the imagination than to enlighten reason—is today almost entirely banished from works of merit. One of our best philosophers seems to have delivered the death blow to it.[43] The spirit of hypothesis and conjecture formerly was perhaps quite useful and even necessary for the renaissance of philosophy, because at that time judiciousness was less important than acquiring independence of thought. But times have changed, and a writer among us who praised systems would have come too late. The advantages now afforded by that spirit are too small to counterbalance the resulting disadvantages, and if the very small number of discoveries they once occasioned are claimed

[42] Buffon (1707–1788), who wrote in a particularly magnificent style, began publishing *l'Histoire naturelle* in 1749. It went through a number of editions in rapid succession and was widely heralded as the great work of the century on all phases of natural history. Over a period of years Buffon continued to produce volume after volume of his natural history, and much of it was incorporated into the *Encyclopedia*. For a comment on the striking parallels between Buffon's theory of biological continuity and d'Alembert's application of the notion of continuity to all thought and phenomena, see n. 59, p. 49.

[43] D'Alembert's note as follows: "M. l'abbé de Condillac, of the Académie Royale des Sciences de Prusse, in his *Traité des Systèmes*.

as proof of the usefulness of systems, one might just as well counsel our geometers to apply themselves to squaring the circle, because the efforts of several mathematicians to do so have given us a few theorems. The spirit of systems [44] is in physics what metaphysics is in geometry.[45] If it may sometimes be required in order to start us on the way, it is almost never capable by itself of leading us to truth. It can glimpse the causes of phenomena when enlightened by the observation of Nature; but it is for calculations to assure, so to speak, the existence of these causes by determining exactly the effects they can produce and by comparing these effects with those revealed to us by experience. Any hypothesis without such a support rarely acquires that degree of certitude which ought always to be sought in the natural sciences, and which is so seldom found in those frivolous conjectures honored by the name of "systems." If all he could have were conjectures of that kind, the principal merit of the physicist would be, properly speaking, to have the spirit of system but never to create one.[46] Thousands of experiments prove how dangerous the use of systems is in the other sciences.

Physics is therefore confined solely to observations and to calculations; medicine to the history of the human body, of its maladies and their remedies; natural history to the detailed description of vegetables, animals, and minerals; chemistry to the composition and experimental decomposition of bodies.

[44] Later editions: "The spirit of system. . . ."

[45] The passage beginning with this sentence and going to the end of the paragraph is taken directly from the "Introduction" to d'Alembert's scientific treatise *Recherches sur la précession des equinoxes, et sur la nutation de l'axe de la terre dans le système Newtonian*, published in 1749 (*Œuvres*, I, 437). It closely parallels the arguments in Condillac's *Traité des systèmes*, published in the same year.

[46] Robert E. Butts, "Rationalism and Modern Science: d'Alembert and the *Esprit Simpliste*," *Bucknell Review*, VIII (1959), 134, makes the following interpretation of this passage:

> The suggestion here is that though a sense of simplicity and systematization should govern the scientist's use of method, simplicity and systematization as methodological postulates should not be generalized into metaphysical propositions regarded as a priori truths.

In a word, all the sciences are confined, as much as possible, to facts and to consequences deduced from them, and do not concede anything to opinion except when they are forced to. I do not speak of geometry, astronomy, and mechanics, which are destined by their nature always to be perfecting themselves.

Men abuse the best things. That philosophic spirit so much in fashion today which tries to comprehend everything and to take nothing for granted extends even into belles-lettres. Some claim that it is even harmful to their progress, and indeed it is difficult to conceal that fact. Our century, which is inclined toward combination and analysis, seems to desire to introduce frigid and didactic discussions into things of sentiment. It is not that the passions and tastes do not have their own sort of logic, but their logic has principles completely different from those of ordinary logic; these principles must be unraveled within us, and it must be confessed that ordinary philosophy is quite unsuited for the task. Totally immersed in the analysis of our perceptions, philosophy disentangles their nuances much more readily when the soul is in a state of tranquillity than when it is in the throes of passions or of the lively sentiments which affect us. In truth, how could it possibly be easy to analyze such feelings as these with precision? We must indeed surrender ourselves to them in order to know them, even though the moment in which the soul is affected by them is the very time when it is least capable of study. It must be admitted, however, that this spirit of discussion has contributed to freeing our literature from blind admiration for the ancients; it has taught us to value in them only the beauties that we would be compelled to admire in the moderns. But it is perhaps also to the same source that we owe that species of metaphysics of the heart which has seized hold of our theaters.[47] While we do not have to banish it entirely, still less are we obliged to let it thus hold sway. This "anatomy of the soul"

---

[47] D'Alembert is probably referring to the minute analysis of various forms of love, usually the beginning stages of tender young love, by Marivaux (1688–1763) in his comedies of the 1720's and 1730's.

has even slipped into our conversations; people make dissertations, they no longer converse; and our societies have lost their principal ornaments—warmth and gaiety.

Thus, let us not be astonished that our literary works are generally inferior to those of the preceding century. One can find the reason for this circumstance in the very efforts we make to surpass our predecessors. Taste and the art of writing make rapid progress in a short time, once the true route is open; hardly does a great genius glimpse the beautiful before he sees it in its entire extent, and imitation of *la belle Nature* seems restricted to certain limits which are reached in only a generation or two at the most, so that for the following generation imitation is all that remains. Yet this next generation is not content with this share: the riches that it has inherited authorize the desire to increase them. It strives to add to what it has received, and it misses the mark while trying to surpass it. Thus, we have at the same time more principles for good judgment, a larger fund of enlightenment, more good judges, and fewer good works. We do not say of a book that it is good, but that it is the book of a man of intelligence. In this manner did the century of Demetrius of Phalerum immediately follow that of Demosthenes, the century of Lucan and of Seneca that of Cicero and of Virgil, and our century that of Louis XIV.[48]

I refer here only to the century in general, for I am far from satirizing a few men of rare merit with whom we live. The physical constitution of the literary world, like that of the material world, follows unavoidable revolutions of which it would be as unjust to complain as it would be of the change of the seasons. Moreover, even as we owe to the century of Pliny the admirable works of Quintilian and of Tacitus, which the preceding generation would perhaps not have been in a position to produce, our own century will leave monuments

---

[48] D'Alembert and Voltaire alike continually looked back upon the age of Louis XIV as a golden period of literature and the arts, from which there had been a marked decline. At the same time they congratulated their own epoch for the victory of the analytic method of philosophy.

to posterity for which it may rightfully glorify itself. A poet, famous for his talents and his misfortunes, has overshadowed Malherbe in his odes and Marot in his epigrams and epistolary verses.[49] We have seen the birth of the sole epic poem which France can set over against those of the Greeks, the Romans, the Italians, the English, and the Spanish. Two illustrious men, among whom our nation seems divided and whom posterity will know how to rank in their proper places, compete for the glory of the buskin, and we see their tragedies with an extreme pleasure even after those of Corneille and of Racine.[50] One of the two men, to whom we owe the *Henriade*, is sure of obtaining a distinguished and most particular place among the very small number of great poets, and at the same time he possesses to the highest degree a talent which hardly any poet has had even to a mediocre degree: that of writing in prose. No one has better known the rare art of presenting each idea effortlessly in the phrase which is proper to it, of embellishing everything without mistaking the proper shading for each thing, and finally, of never being either above or below his subject—a quality that characterizes great writers more than we would imagine. His *Essay on the Century of Louis XIV* [1739] is all the more precious because the author had no model for this type of writing, either from among the ancients or ourselves. By the swiftness and nobility of its style his *History of Charles XII* [1731] is worthy of the hero that he was to portray. His occasional writings, which are su-

---

[49] Jean Baptiste Rousseau (1671–1741) is the poet in question here. He was banished from France because of some satirical couplets he was alleged to have written. Clément Marot was an early-renaissance French poet (1496–1544). For Malherbe see n. 8, p. 67.

[50] Voltaire (1694–1778) and Crébillon (1674–1762). Crébillon was a highly admired tragedian in the classical tradition of the seventeenth century. Of Voltaire's many tragedies, only one or two are still performed on rare occasions; Crébillon has been completely forgotten. Voltaire's *Henriade* (1728), by our standards an excessively long epic poem, helped establish him as the foremost poet of France and was widely hailed as a great work of literature in the eighteenth century.

perior to all those we hold in highest regard, would be sufficient in their number and their merit to immortalize several writers. Would that I could touch on his numerous and admirable works here in order to pay to that rare genius the tribute of praises that he merits, that he has received so many times from his compatriots, from foreigners, and from his enemies, and to which posterity will give the crowning touch when he is no longer able to enjoy it! [51]

These are not our only riches. A judicious writer [Montesquieu], who is as good a citizen as he is a great philosopher,[52]

[51] The *Essai sur le siècle de Louis XIV*, to which d'Alembert refers here, was published at the end of 1739. It should not be confused with the great *Siècle de Louis XIV*, first published in 1751, of which the *Essai* of 1739 constituted most of the first two chapters. (See Voltaire, *Œuvres complètes*, "Avertissement de Beuchot" to the *Siècle de Louis XIV*, XIV, ix.) At the end of the first edition of the *Siècle de Louis XIV*, Voltaire included a passage handsomely eulogizing the *Encyclopedia*, no doubt in part a response to d'Alembert's recent praises of him in the *Preliminary Discourse*. (Quoted by Theodore Besterman, *Voltaire's Correspondence* [Geneva, 1957], XXI, 38–39.) The continued wooing of Voltaire, especially by d'Alembert, at last attracted him to direct participation in the *Encyclopedia* in 1754. This was a major victory for the project. Thereafter the spirited correspondence of Voltaire and d'Alembert gives us valuable knowledge of the day-to-day history of the *Encyclopedia*.

[52] D'Alembert refers to the *Esprit des lois* (1748) by the great political theorist, Montesquieu (1689–1755). Later d'Alembert claimed that he had dared to render Montesquieu justice in the *Preliminary Discourse* at a time when no one had yet publicly raised a voice in France to defend him ("Éloge de Montesquieu," *Encyclopédie*, V [1755], xviii). Montesquieu, duly appreciative, asked Madame du Deffand to thank d'Alembert, and supported her vigorous efforts to have d'Alembert accepted into the Académie Française (*Œuvres de Montesquieu*, ed. A. Masson, III, 1385 and 1475–76). The ideas of Montesquieu are lightly treated in the *Preliminary Discourse*, but beginning in the second volume of the *Encyclopedia* d'Alembert undertook to assure that information drawn from Montesquieu was adequately represented by adding to certain articles by other contributors (e.g., "Capitulaire" and "Cens"). Later he wrote the famous eulogy to Montesquieu at the head of Volume Five of the *Encyclopedia*. Montesquieu's application of the analytical method to the study of politics became canon for the *Encyclopedia*, and eventually most of his work was swallowed up into a multitude of its articles.

has given us a work on the principles of the laws which is disparaged by a few Frenchmen and honored by all Europe.[53] Excellent authors have written history;[54] precise and enlightened minds have probed its meaning. Comedy has acquired a new form which we would be mistaken to reject, since it has added one more pleasure to our lives; and this form of literature was not as unknown to the ancients as some would like to persuade us is the case.[55] Finally, we have several novels which prevent us from regretting those of the last century.

The Fine Arts are not less honored in our nation. If I can believe the enlightened amateurs, our school of painting is the foremost in Europe, and several works of our sculptors would not have been disowned by the ancients. Perhaps of all these arts, music is the one which has made the most progress among us in the last fifteen years. Thanks to the labors of a manly, courageous, and fruitful genius, foreigners who could not bear our "symphonies" are beginning to enjoy them, and the French at last appear to be persuaded that Lully left much to be done in this branch of the arts. In carrying the practice of his art to such a high degree of perfection, M. Rameau has become simultaneously the model and the object of jealousy of a large number of artists, who deprecate him at the same time they are trying to imitate him.[56] But what more specifi-

---

[53] Later editions include the following passage: "a work that will be an immortal monument to the genius and to the virtue of its author, and to the progress of reason in our century, whose middle part will remain a memorable epoch in the history of philosophy."

[54] Later editions add the words: "ancient and modern."

[55] D'Alembert is probably referring to the *comédie larmoyante* (tearful comedy), which was becoming popular in the mid-eighteenth century. It was designed to touch the sentiment with its portrayal of virtue and tenderness and was intended to be a partial fusion of comedy and tragedy. La Chaussée (1692–1754) published a number of such plays.

[56] Rameau (1683–1764) was the most popular and contentious French composer of the eighteenth century. His musical doctrines produced a bitter factional struggle in which both Rousseau and d'Alembert were his enthusiastic partisans and interpreters. In the *Encyclopedia* Rousseau wrote the bulk of the articles on music and d'Alembert added to them frequently with further exposition of Rameau's theories ("Cadence,"

cally distinguishes him is the fact that he has very successfully pondered on the theory of music; that he has been capable of finding the principle of harmony and of melody in the chord root,[57] that by this method he has reduced to more certain and more simple laws a science which was formerly given over to arbitrary rules, or rules dictated by blind experiment. I readily take the opportunity of celebrating this artist-philosopher in a Discourse directed principally to the praise of great men. His merit, which he has forced our century to acknowledge, will be well known only when time has made envy hold its tongue; and his name, which is dear to the most enlightened part of our nation, cannot offend anyone here. But though it might displease some alleged patrons of the arts, a philosopher would certainly be worthy of pity if even in matters of science and taste he did not permit himself to speak the truth.

Such is the wealth that we possess. What an idea of our literary treasures would result if we added to the works of so many great men those of all the scholarly associations which maintain the taste for sciences and letters and to which we owe so many excellent works! Such societies most assuredly are of great advantage in a state, if certain conditions are observed: they should not be multiplied excessively, thus facilitating the entry of an excessive number of mediocre persons;

───────────

"Basse fondamentale"). In 1752 d'Alembert published an edition of Rameau's theories called *Élémens de musique théorique et pratique suivant les principes de M. Rameau*, which was one of the chief means by which Rameau's musical doctrines were spread throughout Europe. A. Oliver in his *The Encyclopedists as Critics of Music* (New York, 1947), p. 112, claims that "to this very day music scholars have realized that the clearest approach to Rameau's theories is to be found in d'Alembert's *Élémens*." The prickly Rameau was nevertheless highly indignant over certain departures of Rousseau and d'Alembert from adherence to all parts of his dogma, and he bitterly attacked them in print.

[57] This is the *basse fondamentale* of Rameau, which is a fictitious bass line consisting of roots of chords occurring in a succession of harmonies. See W. Apel, *Harvard Dictionary of Music* (Cambridge, 1947), p. 288, and the article "Basse fondamentale" in the *Encyclopedia* by Rousseau and d'Alembert.

they should banish all inequalities that might exclude or discourage men who are endowed with talents that will enlighten others. Genius should be the only superiority recognized by them, and within their ranks honor should be the reward for work. Finally, a way should be found so that talents receive their due and are not deprived by intrigue of their just compensation.[58] For let us not be mistaken: we do more harm to the progress of the mind by misplacing such rewards than in suppressing them. To the honor of letters, let us confess that even without the promise of compensation, scholars do yet increase in number. Witness England, a country to which the sciences owe so much, although their government does nothing for them. It is true that the English nation is not neglectful of the sciences, that it even respects them, and this kind of reward, superior to all others, is doubtless the surest means of making the sciences and arts flourish; because while the government distributes offices, it is the public which bestows esteem. Love of letters, a virtue among our neighbors, is still, in truth, only a fashion among ourselves, and perhaps it will never be anything else. But however dangerous might be that mode which produces a hundred proud and ignorant amateurs for every enlightened patron of the arts, perhaps we owe to it the fact that we have not returned to the barbarism into which a multitude of circumstances tends to precipitate us.

One of the chief of these circumstances is that love of false wit and brilliance that protects ignorance, prides itself in it, and sooner or later will extend it everywhere. Ignorance will be the final fruit and ultimate degree of bad taste—yet also its

[58] Here d'Alembert refers to the various royal societies of sciences, literature, the arts, and professions that existed in several of the monarchies of western and central Europe. He was a member of several of them himself, and aspired to belong to others, particularly the most coveted of all, the Académie Française, which he entered in 1754. Ultimately he became its perpetual secretary. In the *Encyclopedia*, appropriately, he wrote the series of articles having to do with the academies of France and all Europe. Intrigue and unfairness did indeed darken the histories of some of these honorary associations, but for the most part they were great instruments of progress in the sciences and the arts.

remedy. For everything has regular revolutions, and the darkness will be like a sort of anarchy, which is most baleful in itself, but sometimes useful in its consequences. However, let us not hope for such a fearful upset. Barbarism lasts for centuries; it seems that it is our natural element; reason and good taste are only passing.[59]

This would perhaps be the place to parry the recent thrusts of an eloquent and philosophical writer [60] who accused the sciences and the arts of corrupting human mores.[61] It would

[59] Throughout the Enlightenment progress is considered probable but not necessarily inevitable, and history is commonly viewed as a flux rather than a continuous progress, as we see in this paragraph.

[60] D'Alembert's note as follows: "M. Rousseau of Geneva, author of the part of the *Encyclopedia* that concerns music, with which we hope that the public will be very well satisfied, has composed a very eloquent discourse to prove that the re-establishment of the sciences and the arts has corrupted human mores. This discourse was awarded a prize in 1750 by the Académie de Dijon, with the highest praises. It was printed in Paris at the beginning of this year [1751] and has done much honor to its author."

[61] In his famous *Discourse on the Sciences and Arts*, also known as his *First Discourse*, which launched his reputation in France, Rousseau already signals the beginning of his split with the main body of the *philosophes*, partly because of his conviction that human sentiment, the source and mainspring of all moral and social progress of mankind, was being slighted in the great preoccupation with the progress of the sciences. D'Alembert himself indicated his concern over the abuse of the analytical method when applied to the expression of sentiments and passions in literature (pp. 96–97), but nonetheless he put his greatest hopes in the progress of analytical empiricism. Rousseau, however, made the human heart and morality his central concern. For him the progress of the mind was only secondary and could even be harmful if not accompanied by moral progress. He moved into an examination of the world of emotional experience and sentiment that laid the foundations of the Romantic movement. His philosophical and personal disagreements with his former encyclopedic colleagues were pathologically intensified by his own mental instability, which led him to imagine a vast conspiracy surrounding him. The open rupture, long in preparation, was brought on by remarks in d'Alembert's article on Geneva in the *Encyclopedia* about the policies of that city toward theatrical performances. Rousseau replied in his *Lettre à d'Alembert* (1758), where he signaled his complete breach with Diderot and made explicit the degree to which he diverged from the encyclopedists in

hardly be appropriate for us to concur with his opinion at the head of a work such as this and, indeed, the worthy man of whom we speak seems himself to have given suffrage to our work by the zeal and the success of his collaboration on it. We will not reproach him for having confused the cultivation of the mind with the abuse that can be made of it, since he would doubtless answer that this abuse is inseparable therefrom. But we would ask him to examine whether the majority of the evils which he attributes to the sciences and to the arts are not due to completely different causes, whose enumeration would be a long and delicate operation. Letters certainly contribute to making society more pleasing; it would be difficult to prove that because of them men are better and virtue is more common. But one can question whether even ethics has that privilege. And furthermore, is it necessary to abolish certain [62] laws because in their name a few crimes are protected whose authors would be punished in a republic of savages? In sum, even assuming that we might be ready to yield a point to the disadvantage of human knowledge, which is far from our intention here, we are even farther from believing that anything would be gained by destroying it. Vices would remain with us, and we would have ignorance in addition.[63]

Let us end this history of the sciences by noting that the different forms of government, which have so much influence on

---

opinions concerning morality and its relationship to religion. See A. M. Wilson, *Diderot*, p. 291ff., and R. Naves, *Voltaire et l'Encyclopédie* (Paris, 1938), pp. 34–50. Rousseau, however, felt honored to be immortalized in the *Preliminary Discourse*, and he did not argue with him over the objections d'Alembert raises here. See *Correspondance générale de Rousseau*, ed. Th. Dufour (Paris, 1924), II, 11–12 (letter of 1751 to d'Alembert).

[62] Later editions: "to abolish laws. . . ."

[63] Later Rousseau specifically refuted this criticism of d'Alembert's in his *Rousseau juge de Jean-Jacques* (1776):

He has obstinately been accused of wanting to destroy the sciences, arts, theatres, and academies, and to plunge the universe into its original barbarism; and he has insisted, on the contrary, upon conserving existing institutions, maintaining that their destruction could only remove the palliatives while leaving the vices [Cited by Charles Frankel, *The Faith of Reason* (New York, 1948), p. 119.]

[men's] minds and on the cultivation of letters, also determine the principal types of knowledge which are to flourish under them, each of these types having its particular merits. In general, there should be more orators, historians, and philosophers in a republic and more poets, theologians, and geometers in a monarchy. This rule is not, however, so absolute that it cannot be altered and modified by an infinite number of causes.

[PART III]

Following the general reflections and views that we have felt we ought to put at the head of this Encyclopedia, *the time has come at last to inform the public more specifically concerning the work which we are presenting to it. The* Prospectus *[November, 1750] by M. Diderot, my colleague, has already been published with this in view. Since it has been received all over Europe with the highest praises, with his permission I shall place it once more before the public here, with the changes and additions which have seemed advisable to us both.*[1]

It cannot be denied that ever since the revival of letters among us, both the general enlightenment which has spread throughout society and that germ of science which imperceptibly disposes minds toward more profound knowledge have been due in part to dictionaries.[2] [The evident utility of works of this sort has made them so common that our task in this day is to justify rather than to praise them. It is contended that by multiplying the aids and the facility for self-instruction, dictionaries will contribute to extinguishing the

---

[1] These revisions were considerable in some places. A few sections were deleted because they repeated something that had appeared earlier in the *Discourse*, and there were several additions and clarifications, largely in answer to criticisms that had been made by the *Journal de Trévoux*. Except for minor changes of a word or phrase, we have indicated all departures from the original version as printed in Diderot's *Œuvres*, ed. Assézat and Tourneux (Paris, 1876), XIII, 129 ff. Two short paragraphs from the original were deleted at the very beginning of the version appearing in the *Discourse*. The first, probably by the publishers, claimed falsely that the work being announced was already finished and ready to be printed. The second, by Diderot, merely announced that he intended to describe the nature of the *Encyclopedia* and how it was composed.

[2] The sections enclosed in brackets were not in the original *Prospectus*. It is impossible to determine with certainty whether d'Alembert composed the added paragraphs or whether both he and Diderot did.

taste for work and study. As for us, we believe there are good grounds for asserting that our indolence and the decadence of good taste should be attributed to the passion for wit and cleverness and to the abuse of philosophy, rather than to the multitude of dictionaries. At the very most such collections serve to provide a little enlightenment for those who would not have had the courage to find it without their aid; but they will never replace books for those who wish to educate themselves. Dictionaries, by their very form, are suitable only to be consulted, and they do not lend themselves to any continuous reading. Were we to learn that some man of letters, desirous of making a thorough study of history, had chosen the *Dictionary* of Moréri to further that aim, we should then concur with the reproach which some would make against us.[3] We should perhaps be more justified in attributing the alleged abuse of which our critics complain to the multiplication of "methods," of "elements," of "abridgments," and of "libraries," [4] if we were not persuaded that one can never commit excess in making accessible the means by which an individual may educate himself.]

[These means would be abridged even further if all man's discoveries in the sciences and the arts up to our time were reduced to a few volumes. This project (even if it comprised the facts of history that are of true utility) would perhaps not be impossible to execute. It should be hoped at least that someone would attempt it; today we claim only to make a rough

[3] The French scholar Moréri published an historical and biographical dictionary in 1674. Although it was filled with errors as well as useful information, it went through a multitude of revisions and editions, and continued to be a popular work of reference through the middle of the eighteenth century.

[4] The reference here is to the spate of popular works in the eighteenth century designed to make the essentials of some art or science understandable to the interested amateur. In their titles they commonly had phrases such as "Method for," "Elements of," "Library of" and "Abridgment of." The *philosophes* considered one of their most honorable functions was to popularize certain progressive ideas, and they did not sneer at "popular" works.

sketch of it. It would free us at last of so many books whose authors have merely copied one another. What ought to secure us against the satirical attack on dictionaries is that the most estimable journalists might likewise be subject to the same ill-founded criticism. Is not their aim essentially to set forth in abridged form what our century is adding to the enlightenment of the preceding centuries? Is it not to teach how to do without unabridged texts and consequently to tear out those thorns which our adversaries like to leave in our path? How much useless reading would we be spared by good abstracts!]

Thus, we have believed it important to have a dictionary which could be consulted on all matters concerning the arts and sciences and which would serve as much to guide those who have the courage to work for the instruction of others as to enlighten those who are only teaching themselves.[5]

Up to this point no one had planned a work so large, or at least no one had carried it through. Leibniz, who of all scholars was the most capable of understanding the difficulties of such a project, hoped that they would be surmounted. However, there were encyclopedias in his time, and Leibniz was not unaware of the fact when he called for one.[6]

Most of these works appeared before the last century, and they were not entirely held in contempt. It was thought that if they did not manifest great genius, they at least showed evidence of some knowledge and application. But what would these encyclopedias be for us? Think of the progress that has been made since their time in the sciences and the arts! Think of the many truths that are unveiled today which were not dreamed of then! True philosophy was in its cradle; the geometry of infinity was not yet in existence; experimental physics

[5] Following this line a paragraph of the original is deleted. It states the intention of the encyclopedists to describe and show the significance of the interdependence of the branches of knowledge and to trace the history of the efforts of the human mind. (*Œuvres*, XIII, 130.)

[6] Leibniz had called for a revision and improvement of the Latin encyclopedia of Alsted of 1630. See Louis de Jaucourt, *Vie de Mr. Leibnitz*, at the beginning of Leibniz's *Theodicée* (Amsterdam, 1734), I, 42.

had scarcely appeared; dialectics did not exist; [7] laws of sound criticism were entirely unknown. The celebrated authors in every genre of whom we have spoken in this Discourse,[8] and their illustrious disciples, either did not exist or had not written anything. Scholars were not animated by the spirit of investigation and of emulation; the spirit of exactitude and method, which is less prolific perhaps, but more choice and rare, had not yet made its conquest of the different parts of literature, and the Academies, whose works have carried the sciences and the arts so far, had not been instituted.

Although the discoveries of great men and the scholarly societies of which we have just spoken proffered subsequently a mighty source for the creation of an encyclopedic dictionary, it must also be admitted that in other respects the prodigious augmentations of materials made such a work considerably more difficult. Yet it is not for us to judge whether the successors of the first encyclopedists showed courage or presumption, and we would leave them all to enjoy their reputations—including the best known among them, Ephraim Chambers, if we did not have particular reasons to weigh his merit.[9]

The *Encyclopedia* of Chambers, which has had such a large number of editions in rapid succession in London and was translated quite recently into Italian, is deserving, we confess, of the honors it has received in England and in foreign

[7] Dialectics is defined in the *Encyclopedia* (IV, 934), as the art of reasoning and of carrying on a disputation with precision.

[8] "Descartes, Boyle, Huyghens, Newton, Leibniz, les Bernoulli, Locke, Bayle, Pascal, Corneille, Racine, Bourdaloue, Bossuet, etc." are listed in the original *Prospectus* (1750), in *Œuvres*, XIII, 131. It is interesting to observe that d'Alembert completely ignores the most direct spiritual precursor of the encyclopedists. Pierre Bayle (1647–1706), the great Huguenot controversialist and scholar of the late seventeenth century, whose *Historical and Critical Dictionary* of 1697 served in many ways as a foundation of the Enlightenment and was certainly a continual inspiration for the encyclopedists. One may speculate that Bayle was considered too obviously heterodox to be mentioned in the introduction for the *Encyclopedia*.

[9] Ephraim Chambers, *Cyclopaedia or an Universal Dictionary of Arts and Sciences* (2 vols.; London, 1728, and numerous subsequent editions).

countries; but perhaps it would never have seen the light of day in English if in our own language there had not already existed the works from which Chambers indiscriminately and imprudently drew most of the things that went into his Dictionary. What, then, would our French compatriots have thought of a translation pure and simple? It would have excited the indignation of scholars and aroused a cry among the public, who would have acquired, under a new and ostentatious title, nothing but information it had already possessed for a long time.

We do not deny Chambers the justice which is due him. He well understood the value of the encyclopedic arrangement, that is to say the chain by which one can descend without interruption from the first principles of an art or science all the way down to its remotest consequences, or ascend from its remotest consequences back up to its first principles, and which allows us to pass imperceptibly from this science or this art to another and, if we may express ourselves thus, to circumnavigate the literary world without losing our way. We agree that the plan and the design of his dictionary are excellent, and if its execution were brought to a certain degree of perfection, it alone would contribute more to the progress of true science than half of the known books. But despite all the obligations we owe to that author and the considerable utility we have derived from his work, we could not avoid observing that much remained to be added to it. Is it truly possible to imagine that everything relating to the sciences and the arts can be included in two folio volumes? The nomenclature of such an extensive material alone would fill up one of them if it were complete. Imagine, therefore, how many articles must have been omitted or truncated in this work.

We are not indulging here in mere conjecture. The entire translation of Chambers has passed before our eyes, and we found a prodigious number of things to be desired in the sciences; in the liberal arts, we found a [single] word where [whole] pages were necessary; and everything was lacking on the subject of the mechanical arts. Chambers read books, but he saw scarcely any artisans; however, there are many things that one learns only in the workshops. Moreover, omissions in an encyclopedic work have a different character from omissions in other kinds of works. An article left out of a common dictionary only makes it imperfect. In an encyclopedia it breaks the enchainment and is harmful to both the form and the substance; and all the art of Ephraim Chambers was required to mitigate the defect.[10]

But without going on at greater length about the English Encyclopedia, we declare that the work of Chambers is not the unique basis on which we have built. A large number of his articles have been redone, and hardly any of the others has been used without addition, correction, or abridgment; Chambers is simply one among the group of authors whom we have consulted in particular.[11] [The praises that were elicited simply by the project of the translation of the English Encyclopedia [12] six years ago would in themselves have been a sufficient motive for us to have recourse to that Encyclopedia, insofar as our own work did not suffer from it.] [13]

---

[10] A sentence is left out here from the original, stating that a work so imperfect for all readers and containing so little new information for French readers would not have been well received in France. (*Œuvres*, XIII, 132.)

[11] A phrase is deleted from the original at this point, stating that the general arrangement of Chambers is the only thing that is common to both that work and the *Encyclopedia*. (*Œuvres*, XIII, 133). This phrase was probably left out because it could be demonstrated that many of the articles of the *Encyclopedia* were copied directly from the work of Chambers, following a practice that was universal in the compilation of works of this nature in the eighteenth century.

[12] *Journal de Trévoux*, May, 1745. D'Alembert reproduced these praises in the preface to the third volume of the *Encyclopedia*, pp. iv–v, where he attempted to refute the major attacks made by the *Journal de Trévoux* on the work. He pointed out the inconsistency of the journal's attacks by calling attention to its earlier enthusiasm for an encyclopedic project.

[13] Three paragraphs, with the exception of one sentence, are deleted here from the original. They describe the tree of knowledge, the debt to Bacon, the foundation of the encyclopedic system in the nature of our knowledge—all of which are treated in considerable detail earlier in the *Discourse*.

[The mathematical part seemed to us most worthy of being retained, but from the considerable changes which have been made therein, it will be possible to judge how much an exact revision of all the parts was necessary.]

[We departed from the English author initially on his genealogical tree of the arts and sciences, for which we believed it necessary to substitute another. That part of our work has been sufficiently expounded above.] It presents to our readers the outline of a work that can only be executed in several folio volumes and which will someday contain all human knowledge.

Faced with such an extensive subject matter, everyone will agree with us on the following reflection: daily experience teaches only too well how difficult it is for an author to present a profound treatment of the science or the art of which he has made a particular study all his life. What man, then, could be so brash and so ignorant in understanding as to undertake single-handedly to treat all the sciences and all the arts? [14]

It thus became clear to us that to support such a great load as the one we would have to carry, it would be necessary to share it; and forthwith we cast our eyes upon a sufficient number of artisans who were competent and well-known for their talents, and scholars well-versed in the particular discipline which was to be their share of the work. To each one we distributed the part that was suited to him.[15] [Some were even in possession of their parts before we took charge of this work.[16] The public will soon see their names, and we are not afraid

---

[14] This sentence is somewhat changed from the original. The following sentence is deleted: "He who claims to know all only shows that he does not know the limits of the human mind." (*Œuvres*, XIII, 134.)

[15] A few phrases are deleted here from the original, concerning the type of people competent to do certain types of articles. (*Œuvres*, XIII, 135.)

[16] This admission demonstrates how much Diderot and d'Alembert were committed to the form of the *Encyclopedia* as it was determined before they became editors.

of reproach on this score.[17]] Thus each one, since he was concerned exclusively with a subject he understood, has been in a position to evaluate soundly what the ancients and moderns have written concerning it and to add knowledge drawn from his own store to the resources they afforded him. No one has invaded the area of someone else or become involved with a subject that he may never have learned. We have enjoyed a more methodical system, greater certainty, greater range, and more detail than most lexicographers. It is true that this plan has reduced the editor's importance to small proportions; but it has added much to the perfection of the work, and we will always feel we have achieved enough glory if the public is satisfied. [In a word, each of our colleagues has made a dictionary of the part with which he is charged, and we have joined all these dictionaries together.]

[We believe we have had good reasons for following the alphabetical arrangement in this work. It has appeared to us to be more convenient and easier for our readers who, when they wish to inform themselves on the meaning of a word, will find it more readily in an alphabetical dictionary than in any other. Had we treated all the sciences separately by making a separate dictionary of each, the alleged confusion of the alphabetical arrangement would have been present in such a new arrangement. In addition, such a system would have been subject to considerable inconveniences because of the large number of words common to different sciences which it would have been necessary to repeat several times or to place at random. On the other hand, if we had treated each science in a separate and continuing discourse which conformed to the order of ideas and not of words, the form of this work would have been still less convenient for most of our readers. They would not have been able to find anything without difficulty. The encyclopedic arrangement of the sciences and the arts would have gained little, and the encyclopedic arrangement of words, or rather of objects through which the sciences

[17] This last sentence was deleted from later editions.

come together and communicate with one another, would have been completely lost. On the contrary, nothing is easier than to satisfy both methods in the plan we have followed, as we have explained above. For the rest, if it had been a matter of producing a separate treatise in the usual form for each science and each art and of merely gathering these different treatises under the title of Encyclopedia, it would have been much more difficult to recruit such a large number of persons for this work. Most of our colleagues would no doubt have preferred to present their work separately rather than to see it merged with a large number of others. Moreover, in following this latter plan, we would have been forced almost entirely to renounce the use of the English Encyclopedia; yet we wanted to use it because of its reputation and the promises of the old *Prospectus* [1745], which was approved by the public and to which we wished to conform.[18] The entire translation of that encyclopedia has been placed in our hands by the publishers who undertook to publish it. We have distributed it among our colleagues, who have preferred to undertake to revise, correct, and augment it, rather than to commit themselves to the task of compiling articles without having any preparatory materials, so to speak. It is true that a great part of this material has been useless to them; but at least it has allowed them to undertake more willingly the work we hoped they would do, work which several would perhaps have refused had they but foreseen how difficult the task would be for them. On the other hand, some of these scholars, who were in possession of their assignment long before we were editors, were already well into their work following the original alphabetical project; consequently, it would have been

[18] See Le Gras, *Diderot et l'Encyclopédie*, and Wilson, *Diderot*, pp. 73 ff., for the history of the *Encyclopedia* before Diderot and d'Alembert took over the leadership of the project. Originally in 1745 it was intended to be a mere translation of the Chambers work with some corrections and additions. This project was announced in a *Prospectus* in the year 1745 and highly praised by the *Journal de Trévoux*. There followed a complicated and sometimes humorous sequence of difficulties and changes of direction before Diderot at last assumed its editorship.

impossible for us to change this project, even if we had been less disposed to approve of it. Finally, we knew or at least we had occasion to believe that there had been no criticism of the alphabetical arrangement to which our model, the English author, had compiled. Thus, everything combined in obliging us to make this work conform to a plan which we would have followed by choice had we been the masters.]

The only operation in our work as editors which presumes some intelligence consists in filling the voids which separate two sciences or two arts and in linking the chain when an article which appears to belong to several areas has not been produced because one colleague has depended on another to write it. But in order that no one charged with one part be held accountable for mistakes that might slip into the sections added by the editors, we will be careful to distinguish these sections by an asterisk.[19] We will scrupulously keep our promise that the work of others will be sacred for us, and we will not fail to consult the author if, in the course of editing, his work appears to us to require some considerable change.

The different writers whose talents we have employed have put the stamp of their particular styles on each article, as well as that of the style proper to the subject matter and the object of their part. A technique of chemistry will not have the same tone as the description of ancient baths and theaters; the operations of a locksmith will not be set forth in the same way as the studies of a theologian on a point of dogma or discipline. Each thing has its coloration, and the various branches of knowledge would become indistinct if they were reduced to a certain uniformity. Purity of style, clarity, and precision are the only qualities which can be common to all the articles, and we hope that they will be in evidence. Any more license than this might result in the monotony and boredom that are almost inseparable from long works, but that

[19] The asterisk was the sign of attribution for Diderot as editor in the body of the *Encyclopedia*. This promise was not consistently carried through, especially in the last ten volumes of the text, which were published secretly.

should be dispelled from this one by the extreme variety of materials.

We have said enough to inform the public concerning the nature of an enterprise in which it has seemed to take interest, the general advantages which will result from it if it is well executed, the success or failure of those who have tried it before us, the extent of its object, the arrangement to which we have complied, the distribution that has been made of each part, and our functions as editors. We turn now to the principal details of its execution.

All of the materials of the *Encyclopedia* can be reduced to three categories: the sciences, the liberal arts, and the mechanical arts. We will begin with matters concerning the sciences and the liberal arts, and we will end with the mechanical arts.

Much has been written on the sciences. The treatises on the liberal arts are multiplied without number; the republic of letters is inundated with them. But how few present true principles! How many others drown them in a superabundance of words or lose them in studied obscurity; how many there are whose authority apparently may not be gainsaid, but in which an error placed beside a truth either discredits the truth or brings credit to itself by this proximity! It would doubtless have been preferable to write less and write better.

Among all the writers, preference has been given to those who are generally recognized to be the best. It is from them that principles have been drawn. To their clear and precise exposition have been joined examples or authorities whose opinions have remained unquestionable. The common practice is to refer to sources or to make citations in a way that is vague, often unreliable, and nearly always confused. As a result, a reader going through the various parts of an article does not know exactly which author he ought to consult on such and such a point, or if he ought to consult all of them—which makes verification a long and painful process. We have tried as much as possible to avoid this inconvenience by citing directly, in the body of the articles, the authors on whose evidence we have relied and by quoting their own text when it is necessary. We have everywhere compared opinions, weighed reasons, and proposed means of doubting or of escaping from doubt; at times we have even settled contested matters.[20] To the best of our ability, we have destroyed errors and prejudices. Above all, we have tried not to multiply and perpetuate them by protecting discarded opinions uncritically or by proscribing accepted opinions without reason. We have not been afraid of going to considerable lengths on a subject when the interest of truth and the importance of a matter demand it, sacrificing pleasing qualities whenever this could not be harmonized with instruction.

[We will make an important remark on definitions at this point. In the general articles on the sciences we have conformed to an established practice in dictionaries and in other works, which prescribes that in treating a science one begins by giving a definition of it. We have also presented the simplest and shortest definition that was possible for us to make. But it should not be thought that the definition of a science, especially an abstract science, can convey the idea of it to those who do not have some initial acquaintance with it. What is a science, indeed, if not a system of rules or facts relating to a certain object? And how can one convey the idea of this system to someone who is absolutely ignorant of what this system includes? Suppose one says that arithmetic is the science of the properties of numbers; does that make it known to someone who does not know anything about it in the first place? Would it be any more useful to his understanding than the definition of the philosopher's stone as "the secret of making gold"? Properly speaking, the definition of a science consists only in the detailed exposition of the things with which this science is concerned, just as the definition of a body is the detailed description of this body itself. According

[20] D'Alembert's articles were models of the kind of scholarly scrupulousness described here. A fine example is the article "Air." Unfortunately, the other encyclopedists, including Diderot, did not always approximate these standards in their articles, as the enemies of the work were quick to point out.

to this principle, it seems to us that what is called "definition" for each science would better be placed at the end than at the beginning of a book which treats it. It would then be the very much reduced result of all the notions which one had acquired. Moreover, what do these definitions for the most part contain, if not vague and abstract expressions whose meaning is often more difficult to establish than that of the science itself? Such are the words *science, number,* and *property* in the definition of arithmetic already cited. Doubtless general terms are necessary, and we have seen in the Discourse what their utility is; but one could define them as an unavoidable abuse of signs, and could call the majority of definitions an abuse, sometimes voluntary and sometimes unavoidable, of general terms. For the rest, we repeat that we have conformed on this point to common practice, because it is not for us to change it, and because the very form of this Dictionary prevents us from doing so. But while humoring prejudices, we have not been afraid to set forth here some ideas which we believe to be sound. Let us continue with the description of our work.][21]

The realm of the sciences and the arts is far removed from the vulgar world; it is a realm in which discoveries are made every day, but many fanciful tales are told about them. It was important to verify the true ones, to forestall those that are false, to establish the points of departure, and thus to facilitate the search for what remains to be discovered. Facts are cited, experiments compared, and methods elaborated only in order to excite genius to open unknown routes, and to advance onward to new discoveries, using the place where great men have ended their careers as the first step. This has been our aim as we have combined the principles of the sciences and the liberal arts with the history of their origin and of their successive advances. And if we have achieved this goal, worthy minds will no longer busy themselves searching for what was known before them. In the future productions of the sciences and lib-

---

[21] D'Alembert is doubtless the author of this passage.

eral arts it will be easy to distinguish between what the inventors drew from their own resources and what they borrowed from their predecessors. Works will be accurately evaluated, and those men who are eager for reputation and devoid of genius, and who brazenly publish old systems as if they were new ideas, will soon be unmasked. But in order to attain these advantages it was necessary to give an appropriately extensive treatment for each subject, to insist on the essential, and to neglect the minutiae. In addition it was necessary to avoid the rather common failing of holding forth at length on what requires only a word, of proving what is in no way contested, and of commenting on what is already clear. We have been neither niggardly nor lavish with explanations. It will be observed that they are necessary wherever we have put them and that they would have been superfluous where we have left them out. We have also been careful not to amass proofs where we thought that a single solid argument would suffice, and we have multiplied them only on occasions when their force depended on their number and their mutual support.

[The articles which concern the elements of the sciences have been worked out with all possible care; they are indeed the foundation and basis of the others. For that reason the elements of a science can be well stated only by those who have gone far beyond them, for they include the system of general principles which extend to the different parts of the science. And in order to know the most effective way of presenting these principles it is necessary to have made a very extensive and varied application of them.] [22]

Such is the list of all the precautions we had to take. Such are the resources upon which we could count; but thanks to the good fortune of our enterprise, so to speak, other resources have fallen to us. They include some manuscripts sent to us by amateurs or submitted by scholars, among whom we will name here M. Formey, perpetual secretary of the Royal Acad-

[22] For a lucid and philosophical expansion of this subject, see d'Alembert's major article "Élémens des sciences," *Encyclopédie*, V, 491ff.

emy of Sciences and Belles-Lettres of Prussia.[23] That illustrious academician had planned a dictionary almost like our own, and he has generously turned over to us the considerable part of it which he had completed, for which we will not fail to pay him due honor. In addition, each artist or scholar who was charged with composing a part of our Dictionary had been accumulating some private research and observations of his own, which he has been willing to publish by this means. Almost all the articles on general and particular grammar will be of that number.[24] We may venture to assert with confidence that no known work will be either as rich or as instructive as ours concerning the rules and usages of the French language, and likewise, indeed, concerning the nature, origin, and philosophical basis of languages in general. Thus, we will impart to the public several literary resources which it has perhaps never known concerning the sciences as well as the liberal arts.

But the obliging help that we have received from all sides will hardly contribute less to the perfection of these two considerable branches of knowledge: we have enjoyed the protection of persons of high station; several scholars have received

[23] Jean Formey (1711–1797), a French Huguenot scholar, son of refugees from the persecutions of the French monarchy after the Revocation of the Edict of Nantes, is a good example of the very important part played by the Huguenots in the background of the Enlightenment and of the *Encyclopedia* specifically. They were journalists, dictionary-makers, popularizers, reformers, and transmitters of ideas from England and the Low Countries. Bayle could be considered the spiritual father of the *Encyclopedia* in many respects. The chevalier de Jaucourt, a Huguenot nobleman who made the completion of the work possible by writing single-handedly approximately one-fourth of the articles, served as a bridge from the scholars of the Huguenot Dispersion to the heart of the *Encyclopedia*.

[24] Du Marsais (1676–1756), a liberal scholar who was considered the foremost grammarian in his time, was the author of the articles on grammar in the first volume. He was alleged to be the author of the important article "Philosophe," but this has been seriously questioned. See H. Dieckmann, *Le Philosophe* (Saint Louis, 1948). He died while the *Encyclopedia* was still coming out, and there is a eulogy to him by d'Alembert at the beginning of volume VII. His work for the *Encyclopedia* was continued by his disciples, Douchet and Beauzée.

us warmly and have sent materials to us, and public libraries, private collections, compendia, portfolios, etc.—all have been opened to us, both by those who cultivate letters and those who love them. A little adroitness and considerable expense have procured what could not be obtained through pure benevolence, and the compensations have almost always calmed the real worries or the simulated alarms of those we had to consult.

We are most especially conscious of our obligations to M. l'abbé Sallier [1685–1761], keeper of the Royal Library. With that politeness which in him belongs to his very nature and which was enlivened still more by the pleasure of favoring a great enterprise, he has permitted us to select everything that can add enlightenment or enhancement to our Encyclopedia from the rich sources of which he is the depositary. He who thus contrives to serve the aim of his prince justifies and even honors the royal choice, and indeed the sciences and the fine arts conjoined cannot do too much to shed luster through their productions upon the reign of a sovereign who favors them. As for us, the spectators of their progress and their historians, we shall concern ourselves solely with transmitting them to posterity. Let us hope that posterity will say, upon opening our Dictionary: such was the state of the sciences and the fine arts then. May the history of the human mind and its productions continue from age to age until the most distant centuries. May the Encyclopedia become a sanctuary, where the knowledge of man is protected from time and from revolutions. Will we not be more than flattered to have laid its foundations? What an advantage would it have been for our fathers and for us, if the works of the ancient peoples, the Egyptians, the Chaldeans, the Greeks, the Romans, etc., had been transmitted in an encyclopedic work, which had also set forth the true principles of their languages? Therefore, let us do for centuries to come what we regret that past centuries did not do for ours. We daresay that if the ancients had carried through that encyclopedia, as they carried through so many other great things, and if that manuscript alone had escaped from the famous Library of Alexandria, it would have been capable of consoling us for the loss of the others.

Such is the substance of what we had to set forth for the public concerning the sciences and the fine arts. The section on the mechanical arts required no fewer details and no less care. Never, perhaps, has there been such an accumulation of difficulties, and, to conquer them, so little help from books. Too much has been written on the sciences; not enough has been written well on the mechanical arts. For what is the scanty information available in the various authors, compared to the extent and richness of the subject? Among those who have treated it, one was not sufficiently informed, and succeeded more in demonstrating the necessity for a better work than in fulfilling the requirements of his subject. Another has only skimmed over the top of the material, treating it as a grammarian and a man of letters rather than as an artisan. A third has more substance, to be sure, and is more workmanlike, but at the same time he is so brief that the operations of the artisans and the description of their machines, a matter in itself capable of filling substantial volumes, occupies only a very small part of his.[25] Chambers added almost nothing to what he translated from our authors. Thus everything impelled us to go directly to the workers.

We approached the most capable of them in Paris and in the realm. We took the trouble of going into their shops, of questioning them, of writing at their dictation, of developing their thoughts and of drawing therefrom the terms peculiar to

[25] The works to which Diderot refers are most probably John Harris (1667?–1719), *Lexicon technicum, or an Universal English Dictionary of Arts and Sciences* (2 vols.; London, 1736), fifth edition; Thomas Corneille (1625–1709), *Le Dictionnaire des arts et des sciences* (2 vols.; Paris, 1731); and Jacques Savary-Desbruslons, *Dictionnaire universal de commerce* (3 vols.; Paris, 1723–1730). I am indebted to Professor Jacques Proust for his advice on this question. His article "La Documentation technique de Diderot dans l'*Encyclopédie*," *Revue d'histoire littéraire de la France*, LVII (1957), 335–52, has the most complete bibliography of sources for Diderot's technical articles. Charles C. Gillispie has published a splendid selection of the plates under the title, *A Diderot Pictorial Encyclopedia of Trades and Industry* (2 vols.; New York, 1959).

their professions, of setting up tables of these terms and of working out definitions for them, of conversing with those from whom we obtained memoranda, and (an almost indispensable precaution) of correcting through long and frequent conversations with others what some of them imperfectly, obscurely, and sometimes unreliably had explained. There are some artisans who are also men of letters, and we would be able to cite them here; but their number would be very small. Most of those who engage in the mechanical arts have embraced them only by necessity and work only by instinct. Hardly a dozen among a thousand can be found who are in a position to express themselves with some clarity upon the instrument they use and the things they manufacture. We have seen some workers who have worked for forty years without knowing anything about their machines. With them, it was necessary to exercise the function in which Socrates gloried, the painful and delicate function of being midwife of the mind, *obstetrix animorum*.[26]

But there are some trades so unusual and some operations so subtle that unless one does the work oneself, unless one operates a machine with one's own hands, and sees the work being created under one's own eyes, it is difficult to speak of it with precision. Thus, several times we had to get possession of the machines, to construct them, and to put a hand to the work. It was necessary to become apprentices, so to speak, and to manufacture some poor objects ourselves in order to learn how to teach others the way good specimens are made.

It is thus that we have become convinced of men's ignorance concerning most of the objects in this life and of the difficulty of overcoming that ignorance. It is thus that we have put ourselves in a position to show that the man of letters who knows his language the best does not know the twentieth part of its words; that, although each art has its own language, this lan-

---

[26] Diderot, himself son of an artisan, knew how to converse with workingmen, and he had a real fascination with the operations of all the crafts. He did most of the arduous work of assembling the notable articles on the arts and crafts.

guage is still very imperfect; that it is through the long-established habit of conversing with one another that the workers make themselves understood, and this is accomplished much more by the repetition of contingent actions than by the use of terms. In a workshop it is the moment that speaks, and not the artisan.

Here is the method which has been followed for each art. We have treated: (1) its material, the places where the material is found, the manner in which it is prepared, its good and bad qualities, its different kinds, the operations through which one makes it pass (whether before using it or while processing it); (2) the principal things that are made from it, and the manner of making them. (3) We have presented the name, the description, and the shape of the tools and machines, in detached pieces and in assembled pieces, and the patterns of casts and of other instruments whose interior design, profiles, etc., it is appropriate to know. (4) We have explained and illustrated the workmanship and the principal operation in one or several plates, where sometimes one sees only the hands of the artisan in action, and sometimes the whole person of the artisan, working at the most important production of his art. (5) We have collected and defined in the most precise way possible the terms peculiar to the art.

But the general lack of experience, both in writing about the arts and in reading things written about them, makes it difficult to explain these things in an intelligible manner. From that problem is born the need for figures. One could demonstrate by a thousand examples that a simple dictionary of definitions, however well it is done, cannot omit illustrations without falling into obscure or vague descriptions; how much more compelling for us, then, was the need of this aid! A glance at the object or at its picture tells more about it than a page of text.

We have sent designers to the workshops. We have made sketches of the machines and of the tools, omitting nothing that could present them distinctly to the viewer. In the case where a machine deserves some details because of its importance and the multitude of its parts, we have passed from the simple to the more complicated. We have begun in a first figure by assembling as many of its elements as can be observed without confusion. In a second figure we see the same elements, with a few others added. Thus we have successively put together the most complicated machine with no difficulty, either for the mind or for the eye. It is sometimes necessary to go from the knowledge of the work to that of the machine and at other times to descend from the knowledge of the machine to that of the work. Some reflections will be found in the article "Art" on the advantages of these methods and on the occasions where it is appropriate to prefer one to the other.

There are some notions which are common to almost all men and reasoned explanation cannot make them clearer than they already are in men's minds. There are also some objects that are so familiar that it would be ridiculous to make figures of them. In the arts one finds still others so complicated that it would be fruitless to make a drawing of them. In the two first cases we have supposed that the reader was not entirely devoid of good sense and experience; in the last case we refer him to the object itself. There is a golden mean in everything, and we have tried to strike it here. If one tried to depict everything and say everything about a single art, the result would be volumes of text and plates. One would never finish if one undertook to make illustrations of all the states through which a piece of iron passes before being transformed into a needle. For the text to follow all the smallest details of the artisan's procedures would clearly be absurd. As for the figures, we have restricted them to the important movements of the worker and to only those phases of the operation which it is very easy to portray and very difficult to explain. We have held ourselves to essential circumstances, to those whose picture, if it is well executed, necessarily results in the knowledge of the other circumstances which one does not see. We have not wished to resemble the man who would set up signposts at each step in a road, for fear that the travelers might get lost. It suffices that they be placed at every point where travelers would be in danger of losing their way.

Moreover, it is workmanship that makes the artisan, and it is not in books at all that one can learn to work by hand. In our Encyclopedia the artisan will find only some views which he would not perhaps ever have had and some observations which he would have made only after several years of work. We will offer to the studious reader, for the satisfaction of his curiosity, what he would have learned by watching an artisan operate, and to the artisan we will offer what one might hope he would learn from the philosopher in order to advance toward perfection.

We have distributed the illustrations and the plates in the sciences and in the liberal arts with the same spirit and the same economy as in the mechanical arts. However, we have not been able to reduce the number of either to less than six hundred. The two volumes that they fill will not be the least interesting part of the work, because of the care we will take to place on the *verso* of a plate the explanation of the other plate facing it, with references to the places in the rest of the Dictionary relating to each figure.[27] The reader opens a volume of the plates; he sees a machine that whets his curiosity; it is, for example, a powder mill or a paper mill, a sugar mill or a silk mill, etc. Opposite it he will read: figure 50, 51, or 60, etc., powder mill, sugar mill, paper mill, silk mill, etc. Following that he will find a succinct explanation of these ma-

[27] Ultimately there were eleven folio volumes of plates published from 1762 to 1772, and a supplementary volume published in 1777. These splendid folio volumes of plates make up one of the most valuable parts of the *Encyclopedia*. They are the best representations of the crafts and techniques that we have from the eighteenth century and, indeed, of all parts of the economy and daily life of the Old Régime. Apparently a substantial share of the plates on the mechanical arts were copied from or based upon designs that had been made by Réaumur for the *Académie des Sciences* and acquired in a questionable manner for the *Encyclopedia*. See Jacques Proust, "La Documentation technique de Diderot dans l'*Encyclopédie*," cited in n. 25, p. 122.

chines with references to the articles "Powder," "Paper," "Silk," etc.

The quality of engraving will correspond to the perfection of the designs, and we hope that the plates of our Encyclopedia will surpass those of the English Dictionary in beauty even as they exceed them in number. Chambers has thirty plates; the old project promises one hundred and twenty of them, and we will provide at least six hundred. It is not astonishing that the arena expands under our feet; it is immense and we do not flatter ourselves in having covered it all.

In spite of the help from the works which we have just mentioned, we declare without hesitation, in the name of our colleagues and in our own, that we will always be disposed to confess our inadequacies and to profit from enlightenment communicated to us. We will accept it with gratitude, and we will conform to it readily, because we are persuaded that the ultimate perfection of an encyclopedia is the work of centuries. It took centuries to make a beginning; it will take centuries to bring it to an end.[28] [Yet we will be satisfied to have contributed to laying the foundations of a useful work.]

We will always have the inner satisfaction of having spared nothing for its success. One of the proofs which we will adduce for this is that there are sections on the sciences and the arts which have been rewritten as many as three times. It behooves us to say, to the honor of the associated publishers, that they have never refused to support whatever could contribute to perfecting anything in this endeavor. One must hope that the combination of so large a number of circumstances, such as the enlightened intelligence of those who have contributed to the work, the help of persons who have interested themselves

---

[28] A very considerable deletion was made here from the original of the following phrase in capital letters: "mais À LA POSTÉRITÉ ET À L'ÊTRE QUI NE MEURT POINT" (To Posterity and to the Being who never dies). The function of the future generations was to perfect and clarify even more the encyclopedic connection of all knowledge. The *Encyclopedia* was but one step in the portentous and continuous encyclopedic effort which would only be achieved by posterity.

in it, and the emulation of the editors and the booksellers, will produce some good effect.

From everything said above, it follows that in the work which we are announcing the sciences and the arts have been treated as if the reader had no preliminary knowledge of them, that what is worth knowing concerning each matter is presented therein, that the articles explain one another, and consequently that the difficulty of nomenclature is nowhere a problem. Whence we infer that, at least one day, this work could take the place of a library of all the types of knowledge for a gentleman and of all types of knowledge except his own particular domain for a professional scholar, that it will expound the true principles of things and will note their relationships, that it will contribute to the certitude and progress of human knowledge, and that by multiplying the number of true scholars, distinguished artisans, and enlightened amateurs, it will contribute new advantages to society as a whole.[29]

All that remains for us is to name the scholars to whom the public owes this work as much as to ourselves.[30] In naming them we will follow as much as possible the encyclopedic arrangement of materials of which they have taken charge. We have done this so that it would not appear that we would

[29] This is the end of the section based on the *Prospectus* of Diderot.

[30] The list of contributors that follows is only for the first volume. It is included with a minimum of notes, chiefly to provide some notion of the range and character of the contributions to the *Encyclopedia*, and because it was part of the original *Preliminary Discourse*. A number of those who appear here either died or disappeared for some other reason from the company of encyclopedists. As enthusiasm built up for the work, however, more and more contributors volunteered their services. Some were eminent writers like Voltaire; others were merely interested scholars from a variety of walks of life who contributed sometimes only one article, sometimes several, on the subjects of their special interest. The total number of contributors rose to nearly two hundred before the seventeen original volumes of the text were completed. The most thorough list of them appears in K. Takeo, *et al.*, "Les Collaborateurs de l'*Encyclopédie*," in *Zinbun*, no. 1 (Kyoto University, 1957), but that is still not exhaustive.

try to assign any distinction of rank and merit among them. The articles of each will be designated in the body of the work by individual letters, the list of which will be found immediately after this Discourse.

We owe *Natural History* to M. Daubenton, Doctor of Medicine, Member of the Royal Academy of Sciences, Curator and Demonstrator of the Cabinet of Natural History, an immense collection, very intelligently and carefully gathered, which cannot fail to be carried to the highest degree of perfection in hands so able as his. M. Daubenton is the worthy colleague of M. de Buffon in the great work on Natural History, whose first three volumes so far published have had three editions in rapid succession. The public awaits its following volumes with impatience. The article *Abeille* which M. Daubenton composed for the Encyclopedia has been published in the Mercure of March, 1751, and its general success led us to insert the article *Agate* in the second volume of the Mercure of June, 1751. This latter has shown that M. Daubenton is able to enrich the Encyclopedia with new and important remarks and views in the part for which he is responsible, just as the article *Abeille* has shown the precision and the clarity with which he is able to present what is known.

Theology is by M. l'abbé Mallet, Doctor of Theology of the Faculty of Paris, of the House and Society of Navarre, and Royal Professor of Theology in Paris. Without any solicitation on his part, his knowledge and his merit alone have caused him to be named to the chair he occupies, which is no small eulogy in our century. M. l'abbé Mallet is also the author of all the articles on *Ancient and Modern History,* a matter in which he is very well versed, as will soon be seen by the important and interesting work he is preparing in this genre. For the rest, it will be apparent that the articles on *History* in our Encyclopedia cannot include the names of kings, scholars, and peoples, which are the particular object of the Dictionary of Moréri, and which would almost have doubled the length of our own. Finally, to M. l'abbé Mallet we owe, in addition, all the articles which have to do with *Poetry, Eloquence,* and

Literature in general. He has already published on this subject two useful works filled with judicious reflections. One is his *Essai sur l'étude des Belles-Lettres* and the other is *Principes pour la lecture des poëtes*. It can be seen from the details we have just given how useful M. l'abbé Mallet has been to this great work by the variety of his knowledge and talents and how much the Encyclopedia is obligated to him. Such obligation could know no excess.

Grammar is by M. Du Marsais, whose name alone is sufficient recommendation.

Metaphysics, Logic, and *Ethics* are by M. l'abbé Yvon, a metaphysician who is profound and, what is even more rare, extremely clear. One can judge this from his articles in this first volume, [among others the article *Agir*, to which we refer, not by preference, but because being short it can show in a moment how sound is the philosophy of M. l'abbé Yvon, how precise and clear his metaphysics].[31] M. l'abbé Pestré, by his knowledge and his merit worthy of seconding M. l'abbé Yvon, has aided him in several articles on Ethics. We take this occasion to announce that M. l'abbé Yvon is preparing conjointly with M. l'abbé de Prades [32] a work on religion, all the more interesting because it will be done by two men of intelligence and two philosophers.

Jurisprudence is by M. Toussaint, Advocate in Parliament and Member of the Royal Academy of Sciences and Belles-

---

[31] The *errata* section of volume II of the *Encyclopedia*, pp. iii–iv, states that this phrase should be deleted, since the article was not written by Yvon but taken from a Jesuit work.

[32] See p. xxvii of the Introduction for mention of the intriguing relationship between the *Preliminary Discourse* and the heretical thesis of the abbé de Prades, defended at the Sorbonne on November 18, 1751, several months following the appearance of the first volume of the *Encyclopedia*. Rumors passed about that Diderot or d'Alembert had written the thesis, or at least inspired it—there were passages in it directly from the *Discourse*—and this statement about a forthcoming theological work of Yvon and de Prades seemed to be damning evidence that the editors had some part in it. D'Alembert claimed to have had nothing to do with it in the introduction to volume III, p. i.

Lettres of Prussia, a title which he owes to the extent of his knowledge and to his talent for writing, which has made a name for him in literature.

Heraldry is by M. Eidous, formerly Engineer of the Armies of his Catholic Majesty, to whom the republic of letters owes the translation of several meritorious works of different kinds.

Arithmetic and *Elementary Geometry* have been reviewed by M. l'abbé de la Chapelle, Royal Censor and Member of the Royal Society of London. By their success his *Institutions de Géométrie* and his *Traité des Sections coniques* have justified the approbation that the Academy of Sciences has given them.

The articles on *Fortification,* on *Tactics* and on the *Military Art* in general are by M. Le Blond, Professor of Mathematics to the Pages of the King's Great Equerry. He is very well known to the public because of his many justly esteemed works, among others his *Elémens de Fortification,* reprinted several times; his *Essai sur la Castramétation;* his *Elémens de la Guerre des Siéges;* and his *Arithmétique et Géometrie de l'Officier,* which the Academy of Sciences has approved with praise.

Stonecutting is by M. Goussier, who is very well versed and knowledgeable in all parts of mathematics and physics, and to whom this work owes many other obligations, as will be seen later.

Gardening and *Hydraulics* are by M. d'Argenville, Councillor of the King in his Councils, Master Ordinary of his Chamber of Accounts at Paris, of the Royal Societies of Sciences of London and Montpellier, and of the Académie des Arcades of Rome. He is the author of a work entitled *Théorie et Pratique du Jardinage, avec un Traité d'Hydraulique,* whose recognized merit and utility are proved by its four editions in Paris and two translations, one into English and the other into German. Since that work deals only with formal gardens and the author considers hydraulics in relation only to gardens, he has generalized these two matters in the Encyclopedia by writing on all kinds of fruit, kitchen, and legume gardens. Further, a new method of pruning trees and new figures of his invention will be included therein. He has also enlarged the part on hydraulics by discussing the finest European machines for raising water, as well as sluices and other constructions that are built in the water. M. d'Argenville is advantageously known to the public for several works in different fields, among others for his *Histoire Naturelle éclaircie dans deux de ses principales parties, la Lithologie et la Conchyliologie.* The success of the first part of that history had led to a second, which will appear shortly and will treat minerals.

Nautical Matters are handled by M. Bellin, Royal Censor and Engineer Ordinary of the Navy, whose labors have produced several maps which scholars and navigators have eagerly welcomed. Our nautical plates will show that this subject is well known to him.

Watchmaking and the *description of astronomical instruments* are by M. J.-B. Le Roy, who is one of the sons of the celebrated M. Julien Le Roy and who combines the instruction received on this subject from a father so esteemed in all Europe with a considerable knowledge of mathematics and physics, and a mind cultivated by the study of belles-lettres.

Anatomy and *Physiology* are by M. Tarin, Doctor of Medicine, whose works on this matter are known and approved by scholars.

Medicine, Medical Material, and *Pharmacy* are by M. Vandenesse, Regent Doctor of the Faculty of Medicine of Paris, who is very well versed in the theory and practice of his art.

Surgery, by M. Louis, Graduate Surgeon, Royal Demonstrator at the Collège de Saint-Côme, and Councillor Commissary for the abstracts of the Royal Academy of Surgery. M. Louis, who although he is very young is already much esteemed by the most able of his colleagues, has been assigned the surgical part of this dictionary by the selection of M. de la Peyronie, to whom surgery owes so much and who has done it and the Encyclopedia a good turn by procuring M. Louis for both of them.

Chemistry is by M. Malouin, Regent Doctor of the Faculty of Medicine in Paris, Royal Censor, and Member of the Royal Academy of Sciences, author of a *Traité de Chimie,* which has had two editions, and of a *Chimie medicinale,* which both Frenchmen and foreigners have greatly appreciated.

Painting, Sculpture, Engraving are by M. Landois, who combines the knowledge of those arts with great intelligence and talent for writing.

Architecture is by M. Blondel, who is an architect celebrated not only for the several works which he has had executed in Paris and for the others for which he has made designs and which have been executed by different sovereigns, but also for his *Traité de la Décoration des Edifices,* whose highly-esteemed plates he engraved himself. We owe to him also the last edition of *Daviler* [33] and three volumes of *l'Architecture Françoise* in six hundred plates. These three volumes will be followed by five others. Love of the public welfare and the desire to contribute to the advancement of the arts of France led him to found a school of architecture in 1744, which in a short time has become very popular. Besides architecture which he teaches his students there, M. Blondel also has able men to teach the parts of mathematics, fortification, perspective, stonecutting, painting and sculpture, etc., that are related to the art of building. For all sorts of reasons, therefore, one could hardly make a better choice for the Encyclopedia.

M. Rousseau of Geneva, of whom we have already spoken and who understands the theory and practice of *Music* as a philosopher and man of intelligence, has contributed the articles that concern that science. A few years ago he published a work entitled *Dissertation sur la Musique moderne.* This work presents a new system of musical notation which might be adopted, perhaps, were it not for the existence of prejudice in favor of the older one.

Besides the scholars whom we have just named there are others who have furnished entire articles—and very important ones—for the Encyclopedia, to whom we will not fail to pay our respects.

M. Le Monnier, of the Royal Academies of Sciences of Paris

---

[33] The "Avertissement" of volume II of the *Encyclopedia,* p. ii, corrects this passage. Blondel did only the plates of this work; the edition itself was by M. Mariette.

and of Berlin and the Royal Society of London, and Ordinary Physician to His Majesty at Saint-Germain-en-Laye, has contributed articles which concern the magnet and electricity, two important matters which he has studied with such success and concerning which he has presented some excellent memoranda to the Academy of Sciences of which he is a member. We have announced in this volume that the articles *Aimant* and *Aiguille aimantée* are entirely by him and we will do the same for those which are by him in other volumes.

M. Cahusac, of the Academy of Belles-Lettres of Montauban, author of *Zenëide,* which the public has seen and applauded so often on the French stage, of the *Fêtes de l'Amour et de l'Hymen,* and of several other works which have had much success in the lyric theater, has contributed the articles *Ballet, Danse, Opera, Decoration,* and several others of shorter length which relate to those four principal ones. We will be careful to designate each of those which we owe to him. In the second volume will be found the article *Ballet,* which he has filled with interesting researches and important observations. We anticipate that in all of them the profound and reasoned study that he has made of the lyric theater will be in evidence.

I [d'Alembert] have written or edited all the articles on *Mathematics* and general *Physics;* I have also supplied some articles in the other parts, but very few in number. I have undertaken, in the articles on *Transcendent Mathematics,* to present the general spirit of methods and to indicate the best works where one can find the most important details for each subject which were not of a nature to enter into this Encyclopedia. I have attempted to explain what seemed to me either not explained with sufficient clarity or not explained at all. Finally I have undertaken to present exact, that is to say simple, metaphysical principles in each matter, as much as it was possible for me to do. One can see an attempt at this in Volume One in the articles *Action, Application, Arithmétique universelle,* etc.

But however extensive my work is, it is much less than that of my colleague M. Diderot. He is the author of the longest and most important part of this Encyclopedia, the part most desired by the public, and I daresay the most difficult to execute; that is the description of the arts. M. Diderot has based it upon memoranda which have been furnished to him by workers and by amateurs whose names will soon be noted, or upon the knowledge which he has drawn personally from the workers, or upon the crafts which he has gone to the trouble of observing and for which on occasion he has had models constructed, in order to study them more at his leisure. To the details of this task, which is immense and which he has carried through with the utmost care, he has added another which is no less extensive: providing a prodigious number of articles which were lacking in the different parts of the Encyclopedia. He has applied himself to this task with a courage worthy of the finest centuries of philosophy, a disinterest which honors letters, and a zeal worthy of recognition by all those who love or cultivate them and in particular by the persons who have co-operated in the work of the Encyclopedia. One will see in this volume what a considerable number of articles this work owes to him. Among these articles there are a large number that are extremely long, such as *Acier, Aiguille, Ardoise, Anatomie, Animal, Agriculture,* etc. The great success of the article *Art,* which he had published separately a few months ago, has encouraged him to give his best efforts to the others; [34] and I believe I can guarantee that they are worthy of being compared with *Art,* although they are on different types of subjects. It is pointless to answer here the unjust criticism of a few gentlemen who, because they are doubtless unaccustomed to anything that requires the least attention, have found the article *Art* too systematic and too metaphysical, as if it were possible that it be otherwise. Any article that has an abstract and general term for its object cannot be well executed without going back to philosophical principles, inevitably somewhat difficult for those who are not in the habit of reflecting.

---

[34] Diderot published the article *Art* early in 1751, along with a pamphlet in response to the unkind implications of the *Journal de Trévoux,* which had just commented upon the *Prospectus.*

For the rest, we should state here that it has been our pleasure to observe a large number of gentlemen who understood this article perfectly. With regard to those who have criticized it, we hope that they have the same reproach to make to us on the articles which have a similar object!

Several other persons, without having contributed entire articles, have given important assistance to the Encyclopedia. We have already spoken in the *Prospectus* and in this Discourse of M. l'abbé Sallier and M. Formey.

M. le comte d'Hérouville de Claye, Lieutenant General of the Armies of the King and Inspector General of the Infantry, whose profound knowledge in the military art does not preclude the successful cultivation of letters and sciences, has communicated some very interesting memoranda on *Mineralogy*, for which he has had models made in relief of the processes of making copper, alum, vitriol, copperas, etc., in fourteen factories. We owe to him also some memoranda on colza, madderwort, etc.

We are obliged to M. Falconet, Consultant Physician of the King, and member of the Royal Academy of Belles-Lettres, who possesses a library as large and as extensive as his knowledge, of which he makes the most estimable use: that of opening it without reserve to scholars. He has given us in that respect all the help that we could wish for. This citizen and man of letters, who combines the most varied erudition with the qualities of a man of intelligence and philosophy, has been willing also to look over some of our articles and to give us useful enlightenment and advice.

M. Dupin, Farmer General, known for his love of letters and of the public welfare, has given us all the elucidation which we require concerning potashes.

M. Morand, who does such honor to the surgery of Paris and to the different academies of which he is a member, has communicated some important observations; they will be found in this volume in the article *Artériotomie*.

MM. de Prades and Yvon, whom we have already praised as they deserve, have furnished several memoranda relating to the *History of Philosophy* and some on *Religion*. M. l'abbé Pestré has also given us some memoranda on *Philosophy* which we will take care to designate in the following volumes.

M. Deslandes, formerly Paymaster of the Navy, has furnished on that matter some important remarks which have been used. The reputation he has acquired by his different works should assure the value of all his productions.

M. Le Romain, Engineer in Chief of the Ile de la Grenade, has given us all the elucidation we required concerning sugar works and several other machines which, as a philosopher and an attentive observer, he has had the occasion to see and to examine on his travels.

M. Venelle, who is very well versed in physics and in chemistry, about which he has presented excellent memoranda to the Academy of Sciences, has furnished some useful and important elucidation concerning *Mineralogy*.

M. Goussier, whose name has already been mentioned in connection with *Stonecutting* and who combines the practice of design with considerable knowledge about mechanics, has given M. Diderot the design of several instruments and their explanation. But he has been particularly concerned with the plates of the Encyclopedia (all of which he has edited and almost all of which he has designed), the making of stringed instruments in general, and the manufacture of the organ, an immense machine which he has represented in detail according to the memoranda of M. Thomas, his associate in this work.

M. Rogeau, an able professor of mathematics, has furnished some materials on *Coining* and several figures which he has designed himself or which he has supervised.

One may well imagine that concerning printing and publication, the associated publishers have themselves provided all the help that we could possibly have desired.

M. Prevost, Inspector of *Glassworks,* has given us some enlightenment on that important art.

The article on *Brewing* has been composed according to a memorandum by M. Longchamp, whose considerable fortune and great aptitude for letters have not removed him from the profession of his forefathers.

M. Buisson, manufacturer from Lyon, and formerly inspector of manufacturers, has contributed memoranda on *Dyeing, Drapery,* the *Manufacture of Fine Fabrics, Silk work, the draw- and milling of Silk, Braidmaking,* etc., and some observations on the arts related to those just mentioned, such as the art of *gilding ingots,* of *beating, drawing,* and *spinning gold and silver,* etc.

M. La Bassée has contributed the articles on *Lacework,* whose details are known only to those who have specifically applied themselves to it.

M. Douet has offered every possible instruction concerning the art of the *Gauze-maker,* which he practices.

M. Barrat, excellent worker in his trade, has assembled and disassembled several times in the presence of M. Diderot that admirable machine, the *stocking loom.*

M. Pichard, merchant and manufacturer of hosiery, has contributed information concerning *Hosiery.*

MM. Bonnet and Laurent, *Silk* workers, have set up and worked a *velvet* frame, etc., and a frame for *brocade,* while M. Diderot observed. Its details will be seen in the article *Velours.*

M. Papillon, a celebrated *Wood Engraver,* has contributed a memorandum on the history and the practice of his art.

M. Fournier, a most capable *Type-Founder,* has done the same for *Type-Founding.*

M. Favre has contributed memoranda about the *Locksmith's Art, Edge-tool Making, Casting of Cannons,* etc., in which he is well informed.

M. Mallet, *Pewterer* from Melun, has left nothing to be desired concerning the knowledge of his art.

M. Hill, an Englishman by origin, has communicated a model of an English *Glassworks* executed in relief, and all its instruments, with the necessary explanations.

MM. de Puisieux, Charpentier, Mabile, and de Vienne have helped M. Diderot in the description of several arts. M. Eidous has composed in their entirety the articles on the *Blacksmith's Trade and Horsemanship*, and M. Arnauld, of Senlis, those that concern *Fishing* and *Hunting*.

Finally, a large number of other well-intentioned persons have instructed M. Diderot on the *Manufacture of Slates*, on *Forging, Founding, Sawmilling, Wire-drawing*, etc. Since most of these persons are absent, we could not use their names without their consent; we will name them if they so wish. There are likewise several others whose names have escaped us. As for those whose help has not been of any use, we believe we are free of the responsibility of naming them.

We are publishing this first volume precisely at the time we have promised it. The second volume is already in the press; we hope that the public will not have to wait for the others, nor the volumes of plates. The exactitude with which we keep our word will depend only upon our lives, our health, and our tranquillity. We declare also, in the name of the associated publishers, that in the event of a second edition, the additions and corrections will be published in a separate volume for those who have purchased the first. The persons who are of assistance to us in the rest of the work will be named at the head of each volume.

That is what we would have to say about this immense collection. It is presented with all that could arouse interest in it: the impatience which has been shown for its appearance, the obstacles which have retarded its publication, the circumstances which have caused us to take over the task, the zeal with which we have applied ourselves to this task as if it had been of our choice, the praises that good citizens have given to the enterprise, the limitless help of all types that we have received, the protection of the government, the feeble as well as powerful enemies who have tried vainly to stifle the work before its birth; finally some authors, who, without connivance or intrigue, expect no other reward for their care and their efforts than the satisfaction of having deserved well of their country. We will not try to compare this Dictionary with others; we recognize cheerfully that they have all been useful to us, and our work does not consist in decrying that of anyone. It is for the reading public to judge us: we believe we ought to distinguish that public which reads from the one which only speaks.

End of Preliminary Discourse

DETAILED EXPLANATION
OF THE SYSTEM OF
HUMAN KNOWLEDGE

and

OBSERVATIONS ON BACON'S
DIVISION OF THE SCIENCES

DETAILED EXPLANATION OF THE SYSTEM OF HUMAN KNOWLEDGE

Physical beings act on the senses. The impressions of these beings stimulate perceptions of them in the understanding. The understanding is concerned with its perceptions in only three ways, according to its three principal faculties: memory, reason, and imagination. Either the understanding makes a pure and simple enumeration of its perceptions through memory, or it examines them, compares them, and digests them by means of reason; or it chooses to imitate them, and reproduce them through imagination. Whence results the apparently rather well-founded general distribution of human knowledge into *history*, which is related to *memory;* into *philosophy*, which emanates from *reason;* and into *poetry*, which arises from *imagination*.

MEMORY, FROM WHICH COMES HISTORY

History concerns *facts*, and facts concern either *God*, or *man*, or *nature*. The facts which concern God belong to *sacred history*. The facts which concern man belong to *civil history*, and the facts which concern nature belong with *natural history*.

HISTORY

I. Sacred. — II. Civil. — III. Natural.

I. *Sacred history* is divided into *sacred* or *ecclesiastical history*. The *history of prophecies*, where the account has preceded the event, is a branch of *sacred history*.

II. *Civil history*, that branch of universal history, *cujus fidei exempla majorum, vicissitudines rerum, fundamenta prudentiae civilis, hominum denique nomen et fama commissa sunt,*

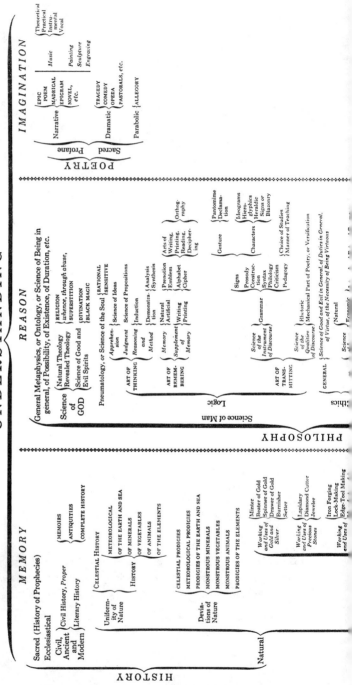

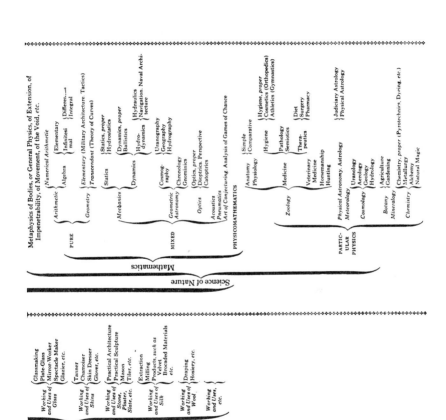

is divided, according to its object, into *civil history* proper and *literary history*.

The sciences are the work of the reflection and of the natural light of men. Chancellor Bacon was therefore justified in saying in his admirable work *De dignitate et augmento scientiarum* that the history of the world without the history of scholars is the statue of Polyphemus with his eye torn out.

Civil history proper can be subdivided into *memoirs, antiquities,* and *complete history*. If it is true that history is the portrayal of past times, *antiquities* are the sketches of it, which are almost always damaged, and *complete history* is a painting, for which *memoirs* are the studies.

III. The divisions of *natural history* derive from the existing diversity of the *facts* of nature, and the diversity of the facts of nature from the diversity of the *states* of nature. Either nature is uniform and follows a regular course, such as one notes generally in *celestial bodies, animals, vegetables,* etc.; or it seems forced and displaced from its ordinary course, as in *monsters;* or it is restrained and put to different uses, as in the *arts*. Nature does everything, either in its *ordinary and regular course,* or in its *deviations,* or in the *way it is employed. Uniformity of nature:* the first part of natural history. *Errors or deviations of nature:* the second part of natural history. *Uses of nature:* the third part of natural history.

It is useless to expand on the advantages of the *history of uniform nature*. But if we are asked what purpose the *history of a monstrous nature* can serve, we will answer: to pass from the prodigies of nature's *deviations* to the marvels of *art;* to lead nature further astray or to put it back on the right road; and above all to temper the boldness of general propositions, *ut axiomatum corrigatur iniquitas.*

As for the *history of nature applied to different uses,* one could make a branch of civil history of it; for art in general is human industry applied to the productions of Nature by virtue of man's needs or his extravagance. Whichever it is, that application is made in only two ways: either in bringing natural bodies together or in separating them. Man can do something or can do nothing according to the possibility or impossibility of joining or separating natural bodies.

The *history of uniform nature* is divided, following its principal objects, into: *celestial history* or *history of the stars,* of *their movements, sensible appearances,* etc., without explaining their cause by systems, hypotheses, etc. (It is a matter here only of pure phenomena.) Into *meteorological history* such as *winds, rains, tempests, thunder, aurora borealis,* etc. Into the *history of the earth and the sea,* or of *mountains, rivers, streams, currents, tides, sands, soils, forests, islands, configurations of the earth, continents,* etc. Into *history of minerals,* into *history of vegetables,* into *history of animals.* Whence results a *history of the elements,* of the *apparent nature, sensible effects, movements,* etc., of *fire, air, earth,* and *water.*

The *history of monstrous nature* should follow the same division. Nature can perform prodigies in the heavens, in the atmosphere, on the surface of the earth, in the interior of the earth, at the bottom of the seas, etc., in everything and everywhere.

The *history of nature applied* is as extensive as the different uses that men make of her productions in the arts, trades, and manufactures. There is no result of the industry of man which cannot be related to some production of nature. The arts of the *minter,* of the *beater, spinner* and *drawer of gold,* of the *burnisher,* etc., can be related to the working of gold and silver and the use to which they are put; the arts of the *lapidary,* the *diamond cutter,* the *jeweler,* the *engraver of precious stones,* etc., to the working of precious stones and the use to which they are put; *iron forging, lock-making, edge-tool making,* the *manufacture of arms,* the *manufacture of small arms, cutlery-making,* etc., to the working of iron and the use to which it is put; *glassmaking, plate-glass making, mirror-making, glazing,* etc., to the working of glass and the use to which it is put; the arts of the *chamois-maker,* the *tanner,* the *skin dresser,* etc., to the working of skins and the use to which they are put; *extracting* and *milling* of wool and silk, the arts of the *drapers, trimming-makers, braid-makers, button-makers, workers in vel-*

vets, satins, damasks, brocaded materials, linenettes, etc., to the working of wool and silk and the uses to which they are put; to the working and use of soil: *clay pottery, faience, porcelain,* etc.; to the working and use of stone: the mechanical functions of the *architect,* the *sculptor,* the *stucco-maker,* etc.; to the working and use of woods: *joinery, carpentry, inlaid work, toy-making,* etc.; and thus with all the other arts, which number more than two hundred. The Preliminary Discourse has shown how much we have proposed to treat each one.

Such is the whole *historical part* of human knowledge that is to be related to the *memory,* and should be the primary material of the philosopher.

REASON, FROM WHICH COMES PHILOSOPHY

Philosophy, or the portion of human knowledge which should be related to reason, is very extensive. There is almost no object perceived by the senses which has not been transformed into a science by reflection. But in the multitude of these objects there are some which stand out by their importance, *quibus abscinditur infinitum,* and to which all the sciences can be related. These principal categories are *God,* to the knowledge of whom man is elevated by reflection on natural and sacred history; *man,* who is certain of his existence through internal sense or consciousness; *nature,* the history of which man has learned through the use of exterior senses. *God, man,* and *nature* will therefore furnish us with a general division of *philosophy* or of *science* (for these words are synonyms); and *philosophy* or *science* will be *science of God, science of man,* and *science of nature.*

PHILOSOPHY OR SCIENCE

I. Science of God. — II. Science of Man. — III. Science of Nature.

I. The natural progress of the human mind is to rise from individuals to species, from species to genera, from closely related genera to distantly related ones, and to create a science at each step; or at least to add a new branch to some science already in existence. Thus the concept, which we meet in history and which sacred history announces to us, of an uncreated and infinite intelligence, etc., and that of the created, finite intelligence united to a body which we observe in man and which we suppose in the brute, have led us to the concept of a created, finite intelligence having no body; and from there to the general notion of the spirit. Moreover, since the general properties of beings, spiritual as well as corporeal, are *existence, possibility, duration, substance, attribute,* etc., these properties have been examined, and from them has been created *ontology,* or the *science of being in general.* Therefore, we have had, in an inverted order, first *ontology;* then the *science of the spirit,* or *pneumatology,* or what is commonly called *particular metaphysics.* And that science is divided into the *science of God* or *natural theology,* which it has pleased God to correct and to sanctify by *Revelation,* whence comes *religion* and *theology proper;* whence through abuse comes *superstition.* Into *doctrine of good and evil spirits,* or of *angels* and of *demons;* whence comes *divination* and the chimera of *black magic.* Into the *science of the soul,* which has been subdivided into *science of the reasonable soul,* which conceives, and *science of the feeling soul,* which is limited to sensations.

II. *Science of Man.* The divisions of the science of man are derived from the divisions of his faculties. The principal faculties of man are the *understanding* and the *will;* the *understanding,* which it is necessary to direct toward *truth;* the *will,* which must be made to conform to *virtue.* The one is the object of *logic,* the other is that of *ethics.*

Logic can be divided into *art of thinking, art of remembering one's thoughts,* and *art of communicating them.*

The *art of thinking* has as many branches as the understanding has principal operations. But four principal operations are differentiated in the understanding: *apprehension, judgment, reasoning,* and *method.* One can relate to *apprehension,* the *doctrine of ideas* or *perceptions;* to *judgment,* that of *propo-*

sitions; to *reasoning* and to *method,* that of *induction* and of *demonstration.*

But in *demonstration,* either one goes back from the thing to be demonstrated to the first principles, or one descends from the first principles to the thing to be demonstrated; whence are born *analysis* and *synthesis.*

The *art of remembering* has two branches, the *science of memory itself,* and the *science of supplements to the memory.* Memory, which we have considered at first as a purely passive faculty and which we are considering here as an active power which reason can perfect is either *natural* or *artificial. Natural memory* is a reaction of the organs, *artificial memory* consists in *prenotion* and in *emblem; prenotion,* without which nothing in particular is present in the mind; *emblem,* by which the *imagination* is called to the aid of the memory.

Artificial representations are the *supplements of memory. Writing* is one of these representations; but one uses either *common characters* or *special characters* in writing. The collection of the first is called the *alphabet;* the others are called *cipher.* Whence are born the arts of *reading,* of *writing,* and of *deciphering,* and the science of *orthography.*

The *art of transmitting* divides into *science of the instrument of discourse* and *science of qualities of discourse.* The science of the instrument of discourse is called *grammar;* the science of the qualities of discourse, *rhetoric.*

Grammar is divided into a science of *signs,* of *pronunciation,* of *construction,* and of *syntax. Signs* are articulated sounds; *pronunciation* or *prosody,* the arts of articulating them; *syntax,* the art of applying them to different intentions of the mind, and *construction,* the knowledge of the order which they should have in discourse, founded on usage and on reflection. But there are signs of thought other than articulated sounds: namely, *gesture* and *characters. Characters* are either *ideograms,* or *hieroglyphics,* or *heraldic signs. Ideograms,* such as those of the Indians, which each indicate an idea and which it is consequently necessary to multiply as many times as there are real beings. *Hieroglyphics,* which are the writing of the world in its infancy. *Heraldic signs,* which form that which we call the *science of heraldry.*

It is thus to the *art of transmitting* that *criticism, pedagogy,* and *philology* should be related. *Criticism,* which restores corrupted parts in the writing of authors, provides editions, etc. *Pedagogy,* which treats of the choice of studies, and of the manner of teaching. *Philology,* which is concerned with the knowledge of universal literature.

It is to the *art of embellishing discourse* that one should relate *versification,* or the *mechanical part* of poetry. We will omit the division of rhetoric into its different parts, because no science or art derives from it, if not perhaps *pantomime,* from gesture; and from gesture and from the voice, *declamation.*

Ethics, which we have made the second part of the *science of man,* is either *general* or *particular.* The latter is divided into *natural, economic,* and *political jurisprudence. Natural jurisprudence* is the science of the duties of man alone; *economic,* the science of the duties of a man as a member of a family; *political,* that of the duties of a man in society. But *ethics* would be incomplete if these factors were not preceded by that of the *reality of moral good and evil;* of the *necessity of fulfilling one's duties,* of being *good, just, virtuous,* etc. Such is the object of *general ethics.*

If one considers that societies are obliged to be virtuous no less than are individuals, one will see the birth of the duties of societies, which could be called the *natural jurisprudence* of a society; *economics* of a society; *interior and exterior commerce of land and sea;* and *politics* of a society.

III. *Science of Nature.* We will divide the science of nature into *physics* and *mathematics.* We again draw this division from reflection and from our tendency to generalize. Through the senses we have received the knowledge of real individual things: *sun, moon, Sirius,* etc., stars; *air, fire, earth, water,* etc., elements; *rain, snow, hail, thunder,* etc., weather; and so forth throughout the rest of natural history. We have at the same time received knowledge of abstractions, *color, sound, taste, odor, density, rarety, heat, coldness, softness, hardness, fluidity, solidity, rigidity, elasticity, weight, lightness,* etc.; *shape, distance, movement, repose, duration, extension, quantity, impenetrability.*

We have observed through reflection that some of these abstractions, such as *extension, movement, impenetrability,* etc., are applicable to all corporeal individuals. We have made them the object of *general physics,* or the metaphysics of bodies; and these same properties, considered in each individual thing in particular, with the varieties which differentiate them, such as *duration, resilience, fluidity,* etc., are the object of *particular physics.*

Quantity, another more general property of bodies which supposes all the others, has formed the object of mathematics. We call *quantity* or *size* everything which can be augmented or diminished.

Quantity, the object of *mathematics,* could be considered either alone and independent of real and abstract individual things from which one gained knowledge of it, or it could be considered in these real and abstract beings, or it could be considered in their effects investigated according to real or supposed causes; this reflection leads to the division of *mathematics* into *pure mathematics, mixed mathematics, physicomathematics.*

Abstract quantity, the object of pure mathematics, is either *numerable* or *extended. Numerable abstract quantity* has become the object of *arithmetic,* and *extended abstract quantity,* that of *geometry.*

Arithmetic is divided into *numerical arithmetic* or *arithmetic by figures,* and *algebra,* or *universal arithmetic by letters,* which is nothing else but the calculation of magnitude in general, and whose operations are properly arithmetical operations indicated in an abridged manner; for, to speak precisely, there is no calculation except by numbers.

Algebra is *elementary,* or *infinitesimal,* according to the nature of the quantities to which one applies it. The *infinitesimal* is either the *differential* or *integral: differential,* if it is a

DETAILED SYSTEM OF HUMAN KNOWLEDGE 153

matter of descending from the expression of a quantity which is finite, or considered as such, to the expression of its instantaneous increment or diminution; *integral,* if it is a matter of returning from that expression to the finite quantity itself.

Geometry either has for its original object the properties of the circle and straight line, or it embraces all types of curves in its speculations; which differentiates it into *elementary* and *transcendent.*

Mixed mathematics has as many divisions and subdivisions as there are real beings in which *quantity* can be considered. *Quantity* considered in bodies insofar as they are mobile or tend to move is the object of *mechanics. Mechanics* has two branches, *statics* and *dynamics. Statics* has for its object *quantity* considered in bodies in equilibrium, and only tending to move. *Dynamics* has for its object *quantity* considered in bodies in motion. *Statics* and *dynamics* each have two parts. *Statics* is divided into *statics proper,* which has for its object *quantity* considered in solid bodies in equilibrium and only tending to move; and into *hydrostatics,* which has for its object *quantity* considered in fluid bodies in equilibrium and only tending to move. *Dynamics* is divided into *dynamics proper,* which has for its object *quantity* considered in bodies in motion; and into *hydrodynamics,* which has for its object *quantity* considered in fluid bodies in motion. But if *quantity* is considered in *waters* that are in motion, *hydrodynamics* takes then the name of *hydraulics. Navigation* can be related to hydrodynamics, and *ballistics* or the shooting of bombs, to mechanics.

Quantity considered in the movements of celestial bodies produces *geometric astronomy;* whence comes *cosmography,* or the *description of the universe,* which is divided into *uranography,* or the *description of the heavens;* into *hydrography* or the *description of waters;* and into *geography;* whence comes further *chronology;* and *gnominics,* or the *art of constructing dials.*

Quantity considered in light produces *optics.* And *quantity* considered in the movement of light produces the different branches of *optics*. Light moving in a direct line, *optics proper;* light reflected in one and the same medium, *catoptrics;* light broken up while passing from one medium to another; *dioptrics*. It is to *optics* that one should relate *perspective*.

Quantity considered in sound, in its vehemence, its movement, its degrees, its reflections, its velocity, etc., gives *acoustics*.

Quantity considered with respect to air, its weight, its movement, its condensation, rarefication, etc., gives *pneumatics*.

Quantity considered in the possibility of events gives the *art of conjecturing;* whence is born the *analysis of games of chance*.

The object of the mathematical sciences being purely intellectual, we should not be surprised at the precision of its divisions.

Particular physics ought to follow the same divisions as natural history. From the history, perceived by the senses, of the *stars,* their *movements, sensible appearances,* etc., reflection has gone on to investigate their origin, the causes of their phenomena, etc., and has produced a science which is called *physical astronomy,* to which one should relate the *science of the influence of the stars,* which is called *astrology;* whence come *physical astrology* and the delusion of *judiciary astrology*. From the history, perceived by the senses, of *winds, rains, hail, thunder,* etc., reflection has gone on to investigate their origins, causes, effects, etc., and has produced the science which is called *meteorology*.

From the history, perceived by the senses, of the *sea, earth, rivers, streams, mountains, tides,* etc., reflection has turned to investigating their causes, origins, etc., and has given birth to *cosmology,* or *science of the universe,* which is divided into *uranology,* or *science of the heavens;* into *aerology,* or *science of the air;* into *geology,* or *science of the continents;* into *hydrology,* or the *science of waters*. From the history of *mines,* perceived by the senses, reflection has gone on to investigate their formation, the working of them, etc., and has produced the science called *mineralogy*. From the history of *plants,* perceived by the senses, reflection has gone on to investigate their economy, propagation, culture, vegetation, etc., and has engendered *botany*, of which *agriculture* and *gardening* are two branches.

From the history of *animals*, perceived by the senses, reflection has gone on to investigate their preservation, propagation, use, organization, etc., and has produced the science which is called *zoology*; whence have come *medicine, veterinary medicine, horsemanship, hunting, fishing,* and *falconry; simple* and *comparative anatomy*. *Medicine* (according to the division of Boerhaave) is occupied either with the constitution of the human body and *reasons* concerning its anatomy, whence is born *physiology;* or it is occupied with the way of preventing illnesses, and is called *hygiene;* or it considers the sick body and treats of causes, of differences, and of symptoms of maladies, and is called *pathology;* or it has for its object the indications of life, of health, of *sicknesses,* their diagnosis and prognosis, and takes the name of *semiotics;* or it teaches the art of healing and is subdivided into *diet, pharmacy,* and *surgery,* the three branches of *therapeutics*.

Hygiene can be considered relative to the *health* of the body, to its *beauty,* or to its *strength,* and is subdivided into *hygiene proper, cosmetics,* and *athletics. Cosmetics* will give *orthopedics,* or the *art of procuring a good conformation for the members of the body;* and *athletics* will produce *gymnastics,* or the *art of exercising them*.

From the experimental knowledge or from the history, perceived by the senses, of *exterior qualities* which are *sensible, apparent,* etc. of *natural bodies,* reflection has led us to the artificial investigation of their interior and hidden properties; and this art is called *chemistry. Chemistry* is the imitator and rival of nature: its object is almost as extensive as that of nature itself; either it *decomposes* beings, or it *re-compounds* them, or it *transforms* them, etc. *Chemistry* has given birth to *alchemy* and to *natural magic. Metallurgy,* or the *general art of treating metals,* is an important branch of *chemistry*. Further, one can relate *dyeing* to that art.

Nature has its deviations, and reason has its abuses. We have related *monsters* to the deviations of nature; and it is to the abuse of reason that one should relate all the sciences and all the arts which show only the greediness, the wickedness, and the superstition of man, and which dishonor him.

Such is the whole of the *philosophic part* of human knowledge which is to be related to reason.

IMAGINATION, FROM WHICH COMES POETRY

History has for its object individual beings, which exist at present or which have existed in the past; and poetry has for its object imaginary individual beings, which are the imitation of historical beings. Therefore it should not be astonishing that poetry follows one of the divisions of history. But the different genres of poetry, and the difference of its subjects, offer us two very natural divisions. The subject of a poem is either *sacred,* or it is *profane;* either the poet speaks of past things, or he makes them present, putting them in action; or he gives body to abstract and intellectual beings. The first of these types of poetry will be *narrative;* the second, *dramatic;* the third, *parabolic.* The *epic poem,* the *madrigal,* the *epigram,* etc. are ordinarily *narrative poetry. Tragedy, comedy, opera, pastoral,* etc., are *dramatic poetry;* and *allegories,* etc., are *parabolic* poetry.

POETRY

I. Narrative. — II. Dramatic. — II. Parabolic.

We mean here by *poetry* only that which is fiction. As there can be versification without poetry, and poetry without versification, we have thought that we should regard *versification* as a quality of style, and consider it as belonging to the art of oratory. On the other hand, we will relate *architecture, music, painting, sculpture, engraving,* etc., to poetry, for it is no less true to say of a painter that he is a poet, than to say of a poet that he is a painter, and of a sculptor or an engraver that he is a painter in relief or in depth, than of a musician that he is a painter through his sounds. The *poet,* the *musician,* the *painter,* the *sculptor,* the *engraver,* etc., imitate and counterfeit nature; but one uses *discourse,* the other, *colors,* the third, *marble, bronze,* etc., and the last, the *musical instrument* or the *voice. Music* is *theoretical* or *practical, instrumental* or *vocal.* With regard to the *architect,* he imitates nature only imperfectly by the symmetry of his works. (*See* Preliminary Discourse.)

Poetry has its monsters as nature does. One should include in this number all the productions of the deranged imagination, and these productions can exist in all the genres.

Such is all the *poetical part* of human knowledge, which one can relate to *imagination,* and the end of our genealogical distribution (or, if you will, our world map) of the sciences and the arts, which we would fear perhaps we had made too detailed, were it not of the most fundamental importance to know well the object of an ENCYCLOPEDIA ourselves, and to expose it clearly to others.

OBSERVATIONS ON BACON'S DIVISION OF THE SCIENCES

I. We have confessed in several places in the *Prospectus* that our *principal obligation* for our encyclopedic tree was to Chancellor Bacon. The eulogy of this great man which was read in the *Prospectus* appears even to have contributed to making known the works of the English philosopher to several persons. Thus, after an avowal so formal, it ought not to be permitted either to accuse us of plagiarism or to try to make us suspected of it.

II. However, this avowal does not preclude a great number of things, especially in the philosophic branch, which we in no way owe to Bacon. It is easy for the reader to evaluate these things. But in order to understand the relation and the difference between the two trees, one must see whether the arrangements are the same. In their material all encyclopedic trees necessarily resemble one another; only the order and arrangement of the branches can differentiate them. One finds virtually the same names of the sciences in the tree of Chambers as in ours; yet nothing could be more different.

III. It is not a question here of the reasons that we have had for following an order other than the one Bacon followed. We have given some of them; it would take too long to explain the others in detail, especially in a matter where the arbitrary could not be completely excluded. Be that as it may, it is for the philosophers, that is for a very small number of people, to judge us on this point.

IV. Some divisions, such as those of mathematics into pure and mixed mathematics, that we have in common with Bacon, are found everywhere, and consequently belong to everyone. Our division of medicine is from Boerhaave, as was announced in the *Prospectus*.

V. Finally, since we have made some changes in the tree of

the *Prospectus,* those who would like to compare that tree with Bacon's should pay attention to these changes.

VI. These are the principles with which one should begin in order to make a somewhat equitable and philosophic comparison of the two trees.

GENERAL SYSTEM OF HUMAN KNOWLEDGE ACCORDING TO CHANCELLOR BACON

General division of human science into *History, Poetry,* and *Philosophy,* according to the three faculties of the understanding: *memory, imagination, reason.*

Bacon observes that this division can also apply to theology. In one part of the Prospectus *this latter idea was followed. But it was abandoned afterward because it appeared more ingenious than solid.*

I.

Division of history into *natural* and *civil*. Natural history is divided into history of the *productions of nature,* the history of the *deviations in nature,* the history of the *uses of nature* or of the *arts.*

Second division of natural history, drawn from *its end* and from *its use,* into *history proper* and *reasoned history.*

Division of productions of nature into *history of celestial things,* of *meteors,* of the *air,* of the *land,* and *sea,* of the *elements,* of the *particular species of individual beings.*

Divisions of civil history into *ecclesiastical, literary,* and *civil history proper.*

First division of civil history proper into *memoirs, antiquities,* and *complete history.*

Division of complete history into *chronicles, lives,* and *relations* [narrations].

Division of the history of the times into *general* and *particular.*

Other division of history of the times into *annals* and *journals.*

Second division of civil history into *pure* and *mixed*.

Division of ecclesiastical history into *particular ecclesiastical history, history of prophecies,* which contains the prophecy and the accomplishment, and *history* of what Bacon calls *Nemesis* or *Providence,* that is to say, of the accord which is sometimes noticed between the revealed will of God and his secret will.

Division of the part of history which is concerned with the *notable sayings* of men into *letters* and *apophthegms*.

II.

Division of Poetry into *narrative, dramatic,* and *parabolic*.

III.

General division of science into *sacred theology* and *philosophy*.

Division of philosophy into *science of God, science of nature, science of man*.

Primary philosophy or *science of axioms,* which extends to all the branches of philosophy. Other branch of that primary philosophy, which treats *transcendent* qualities of beings, *little, much, like, different, being, non-being,* etc.

Science of angels and of spirits, continuation of the science of God, or *natural theology*.

Division of the science of nature or natural philosophy into *speculative* and *practical*.

Division of speculative science of nature into *particular physics* and *metaphysics;* the first having for its object the efficient cause and matter; and metaphysics, the final cause and form.

Division of physics into *science of the principles of things, science of the formation of things* or *of the world,* and *science of the variety of things*.

Division of science of the variety of things into *science of concrete things* and *science of abstract things*.

Division of science of concrete things into the same branches as natural history.

Division of the science of abstractions into the *science of particular properties of bodies*, such as *density, lightness, weight, elasticity, softness*, etc., and *science of movements*, of which Chancellor Bacon makes a rather long enumeration conforming to the ideas of the scholastics.

Branches of speculative philosophy, which consist in *natural problems* and in the *sentiments of ancient philosophers*.

Division of metaphysics into *science of forms* and *science of final causes*.

Division of practical science of nature into *mechanics* and *natural magic*.

Branches of the practical science of nature, which consist in the *enumeration of human, natural or artificial riches* that men enjoy and that they have enjoyed, and the *catalogue of Polychrestes*.

Considerable branch of natural philosophy, speculative as well as practical, called *mathematics*. Division of mathematics into *pure* and *mixed*. Division of pure mathematics into *geometry* and *arithmetic*. Division of mixed mathematics into *perspective, music, astronomy, cosmology, architecture, science of machines*, and several others.

Division of the *science of man* into *science of man proper* and *civil science*.

Division of science of man into *science of the human body* and *science of the human soul*.

Division of the science of the human body into *medicine, cosmetics, athletics*, and *science of pleasures of the senses*.

Division of medicine into three parts, *art of preserving health, art of curing sicknesses, art of prolonging life*. Painting, music, etc., branch of the science of pleasures.

Division of the science of the soul into *science of the divine afflatus* whence came the *reasonable* soul, and the science of the *irrational* soul, which we have in common with the brutes and is produced from the clay of the earth.

Another division of the science of the soul into *science of the substance of the soul, science of its faculties*, and *science of the usage and of the object of its faculties:* from this last result *artificial* and *natural divination,* etc.

Division of faculties of the sensitive soul into *movement* and *feeling.*

Division of the science of the use and the object of the faculties of the soul into *logic* and *ethics.*

Division of logic into *art of inventing,* of *judging,* of *remembering,* and of *communicating.*

Division of the art of inventing into *invention of sciences* or *of arts,* and *invention of arguments.*

Division of the art of judging into *judgment by induction* and *judgment by syllogism.*

Division of the art of syllogism into *analysis* and *principles* for distinguishing easily the true from the false.

Science of analogy, branch of the art of judging.

Division of the art of remembering into *science of that which can aid the memory,* and *science of the memory itself.*

Division of the science of memory into *prenotion* and *emblem.*

Division of the science of communicating into *science of the instrument of discourse, science of the method of discourse,* and *science of the ornaments of discourse,* or *rhetoric.*

Division of the science of the instrument of discourse into *general science of signs* and into *grammar,* which is divided into *science of language* and *science of writing.*

Division of science of signs into *hieroglyphics* and *gestures,* and into *real characters.*

Second division of grammar into *literary* and *philosophic.*

Art of versification and *prosody,* branches of the science of language.

Art of decipering, branch of the art of writing.

Criticism and *pedagogy,* branches of the art of communicating.

Division of ethics into *science of the object* which the soul should have for itself, that is to say of ethical good, and *sci-*

ence of the improvement of the soul. The author makes many divisions in this subject which it is useless to recount.

Division of civil science into *science of conversation, science of business,* and *science of the state.* We omit its divisions.

The author finishes by some reflection on the use of *sacred theology,* which he does not divide into any branches.

Such is the tree of Chancellor Bacon in its natural order and without either dismemberment or mutilation. It can be seen that the category of *logic* is the one where we have followed him most closely, even though we have believed it necessary to make several changes. For the rest, we repeat that it is for the philosophers to judge us on those changes that we have made. Our other readers no doubt will take little part in this question, which was, however, necessary to clarify; and they will only remember the formal avowal that we have made in the *Prospectus* of having our *principal obligation* for our tree to Chancellor Bacon; an avowal which ought to reconcile to us any impartial and disinterested judge.

INDEX [1]

A priori, xxxiii, xxxv, 95
Abbé, title, xx, xliv n
Abstraction, xxxv, 18, 19, 33, 46, 49, 83
Académie Française, xvii, 93 n, 99 n, 102 n
Academies. *See* Scholarly societies
Academy of Sciences, x n, xvii, xxi, xxii, xxviii, 93 n
Aesop, 68
Aesthetics, xxxvii ff., 45, 64 ff., 96, 97
Agriculture, 14, 15, 16, 55
Algebra, 19 n, 20, 78
Analysis, xx, xxxi, xxxii ff., xxxvii, 17 ff., 31, 96, 97, 103 n
Ancients and moderns, quarrel of, 61 n, 96
Anticlericalism, xlvi ff., 71 ff.
Antipodes, 73 f.
Antiquity and ancients, 6, 7, 37, 56, 61 ff., 81, 85, 92, 98, 100, 121
Arabs, 71
Archimedes, 52, 61
Architecture, 37 ff.
Aristotle, xlvi, 71, 80
Arithmetic, 19 ff.
Arrangement or order (alphabetic, encyclopedic, or genealogical), xlix, 45 ff., 76, 110, 113 ff.
Artisans, xlviii, 42, 111, 112, 122 ff.
Arts, 4, 14, 15 n, 30, 32, 34, 40 ff., 103, 104 n
Astronomy, xvi, 21, 35, 73 n, 79 ff.
Attraction, 82

Augustinianism, xlvi
Authority, xxvi, xxxii, xlv ff., 74
Axioms, 6, 27, 28, 44, 84

Bacon, Francis, xi, xxii, xxv, xxx, xxxii ff., xliii, xlix, 25 n, 50, 52 ff., 74 ff., 111 n, 159 ff.
Balzac, Guez de, 66, 67
Barbarism, 6 n, 12, 41, 62 ff., 71, 80, 102, 103, 104 n
Barrow, I., 85
Bayle, P., xliii, 109 n, 120 n
Beaux esprits, 55, 56
Belles-lettres, 60, 64 ff., 68, 70, 93, 96
Benthamites, xl n
Berkeley, Bishop, 9 n
Berthier, Father, x n, xxvi n, xxviii, xxxix n, 4 n
Bertrand, J., xvii, xxv, xxix n
Besterman, T., 99 n
Bodies, 8, 10, 15, 17 ff., 25, 30, 54, 84
Body, human, xxxii, 8, 10 ff., 13, 14, 24, 40, 44, 52, 86
Boerhaave, H., 86, 159
Boileau, 67 n, 68
Boyle, R., 86, 109 n
Budé, 64 n
Buffon, xxiv, 49 n, 50 n, 94, 129
Butts, R., 95 n
Byzantium, 62

Calculus, 81, 85, 86
Canaye, l'abbé de, xxix

[1] For information not cited in this index concerning specific contributors to the *Encyclopedia* and detailed definitions of the various subdivisions of knowledge in the work, the reader should consult the list of contributors, pp. 129 ff., and the exposition of the encyclopedic Chart of Knowledge and of Bacon's system, pp. 143 ff.

Cartesianism and Cartesians, xxxiii, xxxiv ff., 8 n, 9 n, 14 n, 78 ff., 88, 90 ff.
Cassirer, E., xi n, xxxiv, 17 n
Catoptrics, 24
Cause and effect, 9, 21 ff., 82, 95
Certitude, 24, 25, 26, 44, 99
Chambers, *Cyclopaedia* of, xiv, xxv, 109 ff., 122, 127
Chart or Tree of Knowledge, xxx, 48, 55, 57, 143, 144 f.
Chronology, 35
Cicero, 65, 97
Classicism, 37 ff., 67 ff.
Classification, xxxiii, 49
Clergy, xx, xxv, xxxii, xlvii, 67 n, 71 ff.
Clericalism, xxv, xlvi ff., 71 ff.
Color, 18
Columbus, 73
Combining and combination of ideas, xxxvi, 6, 15, 18, 31, 36
Comedy, 68, 96 n, 100
Communication, 11 ff., 30 ff.
Comte, A., 15 n
Condillac, xiii ff., xx, xxiv, xxix, xxxiii ff., xxxv, xli ff., 5 n, 11 n, 22 n, 25 n, 29 n, 33 n, 74 n, 94 n, 95 n
Condorcet, x, xliv, 15 n
Conscience, xl, 44, 45 f., 46 n
Continuity, xxxv ff., 23, 28, 29, 49 n, 94 n, 110
Contributors to the *Encyclopedia*, xxx, 3, 112 ff.; list of, 128 ff.
Corneille, 67, 98, 109 n
Crébillon, 98 n
Curiosity, 16

Deffand, Madame du, ix, xliv n, 99 n
Definition, xxxiv, 117, 118, 123
Democracy and Democratic Ideas, xix, xlvii
Demosthenes, 68, 97
Descartes, xi, xxxii, xli, xlvi, 8 n, 9 n, 14 n, 22 n, 25 n, 77 ff., 90, 109 n
Despréaux. *See* Boileau

Destouches, the chevalier, xv, xvi
Dialectics, 79
Dictionaries, 106 ff.
Diderot, D., xiii ff., xvii ff., xxx, 106 ff., 115 n, 123 n, 134, 135
Dieckmann, H., 32 n, 120 n
Dioptrics, 24, 79
Direct ideas or knowledge, 6, 31, 50
Discovery, 5, 14, 19, 42 ff., 61, 63, 78 ff., 82, 94, 118
Du Marsais, 120 n
Durand, ix n, xiv

Editors of *Encyclopedia*, functions of, 3, 111 ff., 115, 116, 122 ff., 134, 135
Education, xlvii, 31, 34 n, 71, 107 ff.
Electricity, 29
Élémens de musique (d'Alembert, 1752), xvi, 101 n
Élémens de philosophie (d'Alembert, 1759), xxiv, xlix, 49 n
Eloquence, 33, 34, 66
Emotion, xix, 10, 37, 39
Empiricism, xxxii ff., 8 n, 35 n
Encyclopedia of Diderot, history of, xxiv ff.
England, xxii, xxv, xxxiii, xlv, 85 ff., 88, 92, 102, 109, 120 n
Enthusiasm and enthusiasts, 72
Epicureans, 10
Equality, xl, xlvii, 12, 13, 41
Equilibrium, 21, 24, 85
Erudition, 56, 60, 63 ff., 91, 92
Ethics and morality, xxiv, xxvii ff., xxxii, 10 ff., 35, 36, 43, 44, 54, 103 f.
Evidence (as self-evidence), 26, 44
Evil, xxxvii ff., 10 ff.
Existence, 8, 9, 14, 26, 59
Experience, xxxii, 8 ff., 26 ff.
Experiment, xxii, xxxii ff., 24, 25, 75, 79 ff., 95
Extension, 18, 19, 27

Facts, xxxi, xxxii, xxxiv ff., 4, 7, 17 n, 23, 29 n, 75, 96
Faculties of the mind, xxxvi ff., xlii, 32, 50

INDEX

Faith, 72, 73
Fear, xxxviii, 10
Feeling. See Sentiment
Fine Arts, xxxvii, 37 ff., 43, 51, 55 ff., 68 ff., 100
Fontenelle, xxii, xliii, 61 n, 82 n, 93 f.
Formey, xxviii, 119, 120 n
France and the French, xii, xiii, xxi, xxii, xxiii, xxv, 67 n, 73, 77, 88 ff., 98 ff., 102, 110
Francis I, 62
Frankel, C., 104 n
Frederick the Great, ix, xviii, xxix, 89 n
French language, 66 ff., 92, 120

Galileo, 73, 85
Geneva, xiii, xix, 103 n
Genius, 26, 29, 31, 32 n, 34, 42, 43, 45, 51, 60, 61, 64, 67, 75, 76, 79, 81, 85, 88, 91, 100, 102, 118, 119
Geoffrin, Madame, xxii, xliv n
Geography, 35
Geometer, 24, 39 n, 52, 78 ff., 90
Geometry, 19, 21 ff., 27 ff., 46, 51, 52, 78 ff.
Gerbert of Aurillac, 61
Gillispie, C., 122 n
God, xxxix, xlvi, 7 n, 8, 9 n, 13 n, 14, 25, 26, 52 ff., 72, 73, 78, 127 n
Good and highest good, xxxviii ff., 10, 12 ff., 44
Government, xxxviii, 12, 35, 36, 41, 102, 104 f.
Grammar, 33, 43, 120
Gravitation, 79 ff.
Greek language, 62 ff.
Grimm, F. M., 74 n
Grimsley, R., xvii n

Happiness, xlv, 10, 75
Harmony, 70
Harris, J., 122 n
Harvey, 85
Heart, xxxvii, xl, 45, 67, 96, 103 n
Hélvetius, xxxviii, xxxix
Hénault, xlvii
Hippocrates, 75

History, xxxvi, xxxviii, xl, 11 ff., 14 n, 32 ff., 34, 47, 51 ff., 60, 63, 66, 100 n, 118, 129; philosophy of, xxxi, xli ff., 103
Hobbes, xxxviii, 13 n
Holland, xxii, 77, 85 n, 86 n, 88, 120 n
Homer, 52
Horace, 67 n, 68
Hubert, R., 15 n
Huguenots, xxii, 5 n, 9 n, 120 n
Human condition, xi, xii, xxxi, xlvi, 10, 49 n
Huyghens, C., 80, 85
Hypothesis, 7, 22, 24 ff., 28, 79 ff., 94, 95

Ideas, study of, xx, xxxii ff., xxxvi ff., 5 ff., 30 ff., 36 ff., 46 ff., 50, 61, 70, 71
Imagination, xxxvi ff., xlii, 37 ff., 50 ff., 60 n, 66 ff., 70, 76, 86
Imitation, 37 ff., 51, 65 ff., 97
Impenetrability, 17 ff., 27
Innate ideas, xxxii, xl, 6, 7, 80
Inquisition, 73
Instinct, xxxii, xxxvii, 9, 42
Intolerance, 67 n, 72 ff., 80 ff., 120 n
Invention, 42 ff., 79 ff.

Jansenists, 67 n
Jaucourt, chevalier de, xxvii, 38 n, 108 n, 120 n
Jesuits, xxv ff., 34
Journal de Trévoux, x n, xxv ff., xxviii, xxxix n, 4 n, 73 n, 77 n, 106 n, 111 n, 114 n, 135 n
Journal des Sçavans, xxxix n, 9 n, 13 n, 45 n, 54 n
Judgment or validation, xxxvii, xl, 44, 97
Justice and injustice, xxxviii, 12

Kepler, 81

La Bruyère, 65
La Chaussée, 100 n
La Fontaine, 65, 68
La Harpe, J. F., x n

Language, 11, 32 ff., 63 ff., 92, 93, 120, 123, 124
Latin and Latinity, 62 ff., 64 ff., 92, 93
Laws (*see also* Principles), xxxiii, xxxiv, 17 n, 21, 24, 27, 79 ff., 100; civil, xxxviii, 12, 13, 36, 41
Le Brun, 68
Le Gras, J., xxv n, 114 n
Leibniz, xxxv, 86 f., 108, 109 n
Lespinasse, Julie de, xviii
Le Sueur, E., 68
Liberal Arts, 41, 42, 116, 118
Liberty, xxvi, xlv, 62, 72
Linnaeus, 50 n
Locke, J., xx, xxii, xxxii ff., xli, 5 n, 7 n, 11 n, 14 n, 18 n, 25 n, 33 n, 83 ff., 109 n
Logic, 30, 33, 40, 43, 46, 54, 96
Lough, J., lv n
Louis XIV, 67 n, 68 n, 93 n, 97
Louis XV, xxiii, 121
Lovejoy, A. O., xxxv n, 48 n, 49 n
Lucretius, 94
Lully, J., 68, 100

Magnet and magnetism, 23
Malebranche, 9 n, 86
Malherbe, 66, 98
Marivaux 96 n
Marot, 98
Mathematics, xvi, xxxiv, xxxv, 17 ff., 22, 24 ff., 27, 29, 44, 78 ff., 112
Matter, 6 n, 8, 9 n, 13, 18, 20 ff., 27, 86 f. *See also* Bodies
Maupertuis, 89
Mechanical or Manual Arts, 41 ff., 110, 122 ff., 135
Mechanics, 21, 26, 27, 82, 85
Medici, 62
Medicine, 14, 15 n, 16, 24, 55, 75, 95
Mélanges (d'Alembert, 1753, etc.), lv
Memory, xxxvi, xlii, 50 ff.
Metaphysics, xxiv, xxxiv ff., xli, 9, 27, 33, 44, 46, 51 ff., 60, 79 f., 83 ff., 86 f.
Method of philosophy and science, xxxi ff., xxxvii ff., xlv, 16 ff., 25, 31 n, 78 ff., 97 n, 103 n
Michelangelo, 69

Middle Ages, xli, 6, 61 ff., 71
Mind (*esprit*), xxxii, xxxvi ff., xl, 6 ff., 9, 14, 32, 33, 36 ff., 40, 45 n, 50 ff., 60, 61, 70, 76, 86
Molière, 68
Montaigne, 9
Montesquieu, ix, xxii ff., xliii, 3 n, 99 f.
Moréri, 107
Movement, 17 ff., 24, 27 ff., 79 ff.
Muller, M., xvii n
Music, xiv ff., xvii ff., xxiv, 38 ff., 68 ff., 100 f.

Natural law and right, xxiv, xxxvii ff., xl n, 12, 13, 36, 44
Natural or physical science, 15 ff., 70 ff.
Nature, 11, 12, 15, 16, 17, 22, 23, 24, 25, 33, 34, 38, 42, 47, 48, 50, 51, 52, 53, 61 ff., 75, 82, 95; *la belle Nature*, 37, 38, 45, 97
Naves, R., 104 n
Necessity, 11, 14, 16
Needs, human, xxxviii, 10 ff., 14 ff., 25, 29, 38, 43, 46, 47
Newton, I. and Newtonism, xvi, xx, xxii, xxxii, xxxvi, 18 n, 19 n, 79 n, 80 ff., 88 ff., 109

Objects, 8 ff., 33, 37, 40, 70, 75
"Occult qualities," 6, 82
Old Régime, xxi ff., xxvi, xxvii, xliv, xlvii ff., 88 n
Oliver, A. R., xvii n, 39 n, 101 n
Ontology, 53
Optics, 81
Optimism, 10, 87

Pain, xxxviii, 10, 45 n
Painting, 37 ff., 55, 68, 69
Palissot, C., xxix n
Pappas, J. N., xvii n
Pascal, B., x, xl, 45 n, 67 n, 85 f.
Passions, 16, 33, 38, 96, 103 n
Patronage, 62, 63, 66, 78, 101, 102
Persecution. *See* Intolerance
Phaedrus, 68
Phidias, 69

INDEX

Philosophers (*philosophes*), ix, xv, xvi, xviii, xix, xx, xxii ff., xvi ff., xxxii ff., 6 ff., 23, 35 n, 43, 49 n, 55, 56, 61, 64, 71 ff., 90
Philosophy (*philosophie*), ix, xxiv, xxvi ff., xxxii ff., xxxvi ff., xlv ff., 6, 11, 34, 36, 51 ff., 60, 70 ff., 78, 97 n, 108
Physico-mathematical sciences, xvi, xxxiii ff., 21 ff., 26, 27, 78 ff.
Physics, 16, 54, 75 ff.
Picavet, F., x n, lv
Plates for the *Encyclopédie*, ix n, 124 ff., 137
Plato, 94
Pleasure, xxxvii, 10, 16, 37 ff., 43, 56
Pliny, 97
Poetry, 38 ff., 55, 62 ff., 69, 98
Political theory, xix, xxiii, xxxviii, 12, 36, 41, 99
Port-Royal, 67
Posterity, ix, 35, 99, 121, 127 n
Poussin, 68
Poverty, 41, 42
Prades, l'abbé de, xxvi ff., xxxix, 130, 136 f.
Praxiteles, 69
Prejudice, xlvi, 7, 11, 41, 71, 73, 80, 117, 118
Primitive society, 11 ff., 41
Principles or laws, xxxi, xxxiii, xxxiv ff., xlv, 4, 9, 12, 17, 22, 27, 32 ff., 82 ff., 96, 110, 116, 118, 119
Probability, 44
Progress, xi, xii, xxiv, xxx, xxxi, xxxiii, xli ff., 47, 60 ff., 76, 77, 91, 101, 103 n, 104, 121, 128
Properties of Matter, 17 ff., 54
Prospectus of the *Encyclopédie* (1745), 111, 114
Prospectus of the *Encyclopédie* (Diderot, 1750), xxv, xxvi, xxix, xxx, 3, 55, 76, 106 ff., 109 n, 135 n, 159, 164
Proust, J., xxv n, 122 n, 126 n
Psychology, xx, xxxiii ff., xlii
Puget, 68

Quinault, 68
Quintilian, 97

Racine, 65, 67, 98, 109
Rameau, xvi, xix, 68 n, 100 f.
Raphael, 69
Rationalism, xxxii ff., 5
Reason, xxxi, xxxii, xxxvi ff., xlii, 9, 30, 35, 45 n, 50 ff., 60, 63, 70, 76, 100 n
Réaumur, 126 n
Reflection and reflective ideas, 6, 8, 14, 31, 37, 50
Reform, xiii, xxi, xl n, xliii, xliv, xlviii
Régis, P., 91
Religion and Christianity, xxvi, xxxii, xxxix, xlv ff., 26, 67 n, 71 ff., 86
Renaissance, xxx, xli ff., 6, 60 ff., 76, 94
Republic of letters, xii, xiii, xxii, xxiii, xxvii, 3, 116
Revelation, xxiv, xlvi, 26, 52 ff.
Revolution, xi, xii, xix, xliv, xlvi, 35, 62, 80
Rezler, M., xvii n
Rhetoric, 34, 65
Rights, xxiv, 12, 41
Romanticism, xix
Ronsard, 66
Rouhault, J., 91
Rousseau, J. (poet), 98
Rousseau, J.-J., xiii ff., xix, xl, xlvii, 39 n, 100 n, 103 f., 133; *Discourse on the Sciences and Arts*, xix, xxiv, 103
Rulhière, xxiii, 39 n

Sabatier de Castres, x n
Saint-Lambert, 32 n
Sallier, l'abbé de, 121
Salons, xv, xvi, xxi, xxii
Saurin, J., 82
Savary, J., 122 n
Scholarly societies, x n, xii, xvii, xxix, 101 f., 109

Scholars, 42, 48, 55, 56, 63 ff., 92, 93, 119 ff., 128
Scholasticism, xxiv, xlvi ff., **6, 34,** 62 ff., 71 ff., 76, 80, 82, 88
Science, definition of, 4 n, 40, 117
Sciences, 15 n, 32, 36, 42, 70 ff., 103, 104, 119
Sculpture, 37 ff., 69
Secularism, xxxi ff., xlv
Self-preservation, xxxviii, 12, 15, 44
Seneca, 97
Sensation and senses, xxxii ff., 6 ff., 31
Sentiment or feeling, xxxvii, 10, 33, 34, 37, 38, 44 ff., 89, 96, 100, 103 n
Shape, 18 ff.
Signs, 11, 28, 32 ff., 47, 51, 83
Simplicity, xxiv ff.
Skeptics and skepticism, 9
Social contract, 41
Society and social ideas, xxii, xxxviii ff., xli ff., 11 ff., 32, 35, 41, 54, 61, 104
Society for Early Modern Studies, Davis, lv
Socrates, 26, 123
Sorbonne, xxiv, xxvi ff., xliv n
Soul (*âme*), xxxiii, 6, 10, 11, 13, 38, 39, 40, 46, 50, 52, 54, 63, 84, 86, 87, 89, 96
Space, 17 ff.
Stoics, 10
"Substantial forms," 6
Superstition, 69, 76
Sydenham, 86
Synthesis, xxxi, 17, 31
System, xxi ff., 5, 22, 29, 46 ff., 76, 77, 79 ff.; "Spirit of System," xxxv, 22 f., 75, 86, 87, 94, 95, 119

Takeo, K., 128 n
Taste (*goût*), xxiv, xxxvii, 39, 45, 66, 67 ff., 70, 96, 97, 102 f., 107
Technology, xlviii
Tencin, Madame de, xv
Theology, xxiv, xxv, xxxi, xlvi, 53 f., 71 ff.
Thorndike, L., 74 n
Tolerance. *See* Intolerance *and* Liberty
Torrey, N. C., xi n
Tournefort, 50 n
Trades and crafts, xlvii, 122 ff.
Tragedy, 67, 98
Traité de dynamique (d'Alembert, 1743), xvi, 17 n, 18 n, 22, 27
Tree of Knowledge, 20 n, 46 ff., 48, 76, 77, 112, 144 f. *See also* Chart of Knowledge
Truth, xxxii, xl, xlv, 6, 7, 14, 26, 29, 37, 45, 61, 67, 71, 73, 79 ff., 95
Turgot, xxiv, xliii, xliv

Unity, xxxiv ff., 5, 29, 49
Utilitarianism, xxxii, xxxviii ff., xl n, 11
Utility, 12, 15, 16, 42, 75, 107

Vartanian, A., 22 n, 91 n
Venturi, F., xxv n, xxvii n
Vesalius, 86
Viète, F., 78
Virgil, 65, 97
Voltaire, x n, xvii, xxii, xxiii, xliii, 3 n, 10, 68 n, 74 n, 82 n, 97 n, 98, 99 n
Vortices, Descartes', 79 ff., 88, 90

Will, 6, 10, 13, 104
Wilson, A. M., xviii n, xxiii n, xxv n, xxvi n, xxviii n, lv n, 114 n

Zachary, Pope, 73

The Library of Liberal Arts

AESCHYLUS, *Prometheus Bound*, 24
ANTHOLOGIES ON RELIGION
  Ancient Roman Religion,
    ed. F. C. Grant, 138
  Buddhism, ed. C. H. Hamilton, 133
  Hellenistic Religions,
    ed. F. C. Grant, 134
  Islam, ed. A. Jeffery, 137
  Judaism, ed. S. W. Baron and
    J. L. Blau, 135
  *Religions of the Ancient Near
    East*, ed. I. Mendelsohn, 136
ARISTOTLE, *Nicomachean Ethics*, 75
  On Poetry and Music, 6
  On Poetry and Style, 68
ST. AUGUSTINE, *On Christian
    Doctrine*, 80
AVERROES, *On the Relation of
    Philosophy and Theology*, 121
BACON, F., *The New Organon*, 97
BECCARIA, *Of Crimes and
    Punishments*, 107
BERGSON, H., *Introduction to
    Metaphysics*, 10
BERKELEY, G., *New Theory of
    Vision*, 83
  *Principles of Human
    Knowledge*, 53
  Three Dialogues, 39
BOCCACCIO, *On Poetry*, 82
BOETHIUS, *The Consolation of
    Philosophy*, 86
ST. BONAVENTURA, *The Mind's Road
    to God*, 32
BOWMAN, A., *The Absurdity of
    Christianity*, 56
BRADLEY, F. H., *Ethical Studies*, 28
*British Constitutional Law,
    Significant Cases*,
    ed. C. G. Post, 66
BURKE, E., *Appeal from the New
    to the Old Whigs*, 130
  *Reflections on the Revolution
    in France*, 46
BUTLER, J., *Five Sermons*, 21
CALVIN, J., *On the Christian Faith*, 93
  On God and Political Duty, 23
CATULLUS, *Odi et Amo. Complete
    Poetry*, 114

CICERO, *On the Commonwealth*, 111
Cid, The Epic of the, 77
DANTE, *On World-Government*, 15
DESCARTES, R., *Discourse on
    Method*, 19
  Discourse on Method and
    Meditations, 89
  Meditations, 29
  Philosophical Essays, 99
  *Rules for the Direction of
    the Mind*, 129
DEWEY, J., *On Experience, Nature,
    and Freedom*, 41
DOSTOEVSKI, F., *The Grand
    Inquisitor*, 63
DUCASSE, C. J., *Art, the Critics,
    and You*, 81
EMERSON, R. W., *Nature*, 2
EPICTETUS, *The Enchiridion*, 8
EPICURUS, *The Epicurean Letters*, 141
ERASMUS, D., *Ten Colloquies*, 48
EURIPIDES, *Electra*, 26
FICHTE, J. G., *Vocation of Man*, 50
GOETHE, J. W., *Faust I*, 33
GROTIUS, H., *Prolegomena to The
    Law of War and Peace*, 65
HANSLICK, E., *The Beautiful
    in Music*, 45
HARRINGTON, J., *Political
    Writings*, 38
HEGEL, G. W. F., *Reason in
    History*, 35
HENDEL, C. W., *The Philosophy of
    David Hume*, 116
HERDER, J. G., *God: Some
    Conversations*, 140
HESIOD, *Theogony*, 36
HOBBES, T., *Leviathan I and II*, 69
HUME, D., *Inquiry Concerning
    Human Understanding*, 49
  Political Essays, 34
  Principles of Morals, 62
KANT, I., *Critique of Practical
    Reason*, 52
  *Foundations of the Metaphysics
    of Morals*, 113
  *Fundamental Principles of the
    Metaphysic of Morals*, 16

KANT *(cont.)*, *Metaphysical Elements of Justice*, Part I of *Metaphysik der Sitten*, 72
Metaphysical Elements of Virtue, Part II of *Metaphysik der Sitten*, 85
Perpetual Peace (Beck tr.), 54
Perpetual Peace (Smith tr.), 3
Prolegomena to Any Future Metaphysics, 27

KLEIST, *The Prince of Homburg*, 60

LAO TZU, *Complete Works*, 139

Lazarillo de Tormes, The Life of, 37

LESSING, G., *Laocoön*, 78

LOCKE, J., *A Letter Concerning Toleration*, 22
Second Treatise of Government, 31

LONGINUS, *On Great Writing*, 79

LUCRETIUS, *On the Nature of Things*, 142

MACHIAVELLI, N., *Mandragola*, 58

MILL, J., *An Essay on Government*, 47

MILL, J. S., *Autobiography*, 91
Considerations on Representative Government, 71
On Liberty, 61
Nature and Utility of Religion, 81
Theism, 64
Utilitarianism, 1

MOLIÈRE, *Tartuffe*, 87

MONTESQUIEU, *Persian Letters*, 131

NIETZSCHE, F. W., *Use and Abuse of History*, 11

NOVALIS, *Hymns to the Night*, 115

PAINE, T., *The Age of Reason I*, 5

PLATO, *Epistles*, 122
Euthyphro, Apology, Crito, 4
Gorgias, 20
Meno, 12
Phaedo, 30
Phaedrus, 40
Protagoras, 59
Statesman, 57
Symposium, 7
Theaetetus (Cornford tr.), 105
Theaetetus (Jowett tr.), 13
Timaeus (Cornford tr.), 106
Timaeus (Jowett tr.), 14

SEVEN COMMENTARIES ON PLATO'S DIALOGUES
BLUCK, R. S., *Plato's Phaedo*, 110
CORNFORD, F. M., *Plato and Parmenides*, 102
Plato's Cosmology, 101
Plato's Theory of Knowledge, 100
HACKFORTH, R., *Plato's Examination of Pleasure*, 118
Plato's Phaedo, 120
Plato's Phaedrus, 119

PLAUTUS, *The Haunted House*, 42
The Menaechmi, 17
The Rope, 43

ROSENMEYER, T., M. OSTWALD, AND J. HALPORN, *Meters of Greek and Roman Poetry*, 126

SCHOPENHAUER, A., *Freedom of the Will*, 70

SENECA, *Medea*, 55
Oedipus, 44
Thyestes, 76

SMITH, A., *The Wealth of Nations*, 125

SOPHOCLES, *Electra*, 25

SPINOZA, B., *Improvement of the Understanding*, 67

TERENCE, *Phormio*, 95
The Brothers (Adelphoe), 112
The Mother-in-Law, 123
The Woman of Andros, 18

TOLSTOY, L. N., *What Is Art?* 51

VOLTAIRE, *Philosophical Letters*, 124

WHITEHEAD, A. N., *Interpretation of Science*, 117

WHITMAN, W., *Democratic Vistas*, 9